EVEREST REVEALED

EDITED BY CHRISTOPHER NORTON

EVEREST REVEALED

THE PRIVATE DIARIES AND SKETCHES OF EDWARD NORTON, 1922–24

The
History
Press

COVER IMAGES:
Front, from top: Mount Everest from Pang La; Norton sketching.
Back, clockwise from top left: The Pirate King; Kyang and gazelle; 1924 exhibition members at base camp; Meconopsis or Blue Poppy.

FRONTISPIECE:
Fig. 1   Norton sketching Chomolhari from Chu ké, near Phari, 8 April 1922. The sketch is reproduced as Pl. 4.

First published 2014

The History Press
The Mill, Brimscombe Port
Stroud, Gloucestershire, GL5 2QG
www.thehistorypress.co.uk

Original texts and sketches by E.F. Norton © Norton Everest Archive
Additional material © Christopher Norton 2014

The right of Christopher Norton to be identified as the Author
of this work has been asserted in accordance with the
Copyright, Designs and Patents Act 1988.

British Library Cataloguing in Publication Data.
A catalogue record for this book is available from the British Library.

ISBN 978 0 7509 5585 0

Typesetting and origination by The History Press
Printed in India

CONTENTS

PREFACE

This book reproduces diaries, letters and sketches from the hand of Edward ('Teddy') Norton, my grandfather. He was a member of the 1922 and 1924 Mount Everest expeditions, and leader of the latter. On each expedition he set world altitude records for climbing without supplementary oxygen, and he is generally acknowledged to have been one of the finest Everest expedition leaders.

This material is almost all previously unpublished. It may seem surprising to choose this moment, so long after the event, to publish it. The reason is fairly simple. During his lifetime, my grandfather, a modest man, never contemplated publishing either the diaries or the sketches. Indeed, he had to be pushed hard to show them even to close members of the family, claiming that they were of little interest to anyone except himself. Joyce, his widow – as she was for many years – took a similar line, only breaching it once when she consented to the Alpine Club publishing a small selection of the sketches in the *Alpine Journal* (1993). One or two have since been published elsewhere. However, the early Everest expeditions have continued to attract a remarkable degree of public interest, further fuelled by the discovery of George Mallory's body on Everest in 1999. So it seems appropriate, in this 90th anniversary year of the 1924 expedition, to make this material more widely available for the light it sheds not just on the attempts on the mountain itself, but also on the expeditions as a whole. The sketches in particular provide a vivid picture in colour of a world which is generally seen through the lens of black-and-white photography.

I never knew my grandfather, who died before I was born. In preparing this material for publication on behalf of the family, I have been able to draw on the knowledge and memories of my father, Dick Norton, and my uncles Bill Norton and Hugh Norton. They have provided every assistance and have contributed the written sketch of their father at the start of this book. My Uncle Hugh, in particular, has discussed every detail of the volume with me and contributed to its narrative sections. He has also allowed me to read his biographical memoir of his father, which it is hoped to publish separately, and has saved me from a number of errors. I have also benefited from the advice of my brother, Richard Norton, and my cousin Mark Norton. My daughter Anne has provided much technological help and has chased up material for her digitally challenged father on the internet. Edward, my son, has inherited his great-grandfather's love of mountains. Last but by no means least, my wife Sue, who married into the Everest story, has been a constant support. In short, this book is very much a family initiative, and if it was their decision that my name should appear on the title page, I could not have completed (or indeed begun) it without their support.

A particular debt of gratitude is owed to Julie Summers, great-niece of Andrew Irvine and his biographer, who had previously transcribed the text of the diaries while working on Irvine's biography. Her transcription has been used as the basis for the present edition, and she has helped in other ways as well. I also recall her enthusiastic support on a memorable day in June 2012 when descendants of some of the early Everest pioneers had the privilege of meeting the Dalai Lama at Westminster Abbey. On that occasion I had the honour of presenting to the Dalai Lama an album of photographs of some of my grandfather's watercolours of pre-war Tibet.

For help with research on the photographs from the 1922 and 1924 Everest expeditions I am indebted to Audrey Salkeld and David Somervell, who has kindly allowed me to use his father's photographs taken during the summit attempts in 1922 and 1924. The black-and-white photographs are from the Royal Geographical Society (RGS) Mount Everest collections and from the John Noel Photographic Collection, which is deposited with the RGS. I am indebted to Sandra Noel for permitting the use of her father's photographs, and to the staff of the RGS photographic library, Jamie Owen and Joy Wheeler, for much assistance. The text has been typed and retyped by Brittany Scowcroft and Karen Brett. The sketchbooks have been photographed by Paul Sheils and the maps have been drawn by Amanda Daw. I am most grateful to them all for the care that they have given to these tasks.

✳ ✳ ✳

The watercolours and drawings reproduced in this book are contained in three sketchbooks, one from the 1922 Everest expedition and two from 1924. There are also two sketches on loose sheets (Pls 71–2). The pages of the 1922 sketchbook and one of the 1924 sketchbooks measure 10in x 7in. The other 1924 sketchbook is smaller, 7in x 5in. This contains the majority of the drawings of people, animals and birds, whereas the large sketchbooks are predominantly landscapes. There was once a second 1922 sketchbook. This can be deduced from the mentions in the 1922 diary of a number of landscapes which do not survive, and is corroborated by a letter from Norton to the Secretary of the Mount Everest Committee, Arthur Hinks, dated 4 December 1922, in which he mentions *two* sketchbooks. There is no memory within the family of there ever having existed a second 1922 sketchbook, and it appears that it went missing at an early date. It seems likely that it included the kind of portrait sketches of people and animals found in the smaller 1924 sketchbook.

The sketchbooks contain a mixture of cream-coloured and grey pages. The pencil sketches of people, animals and birds, sometimes enhanced with colour wash, are for the most part on the cream pages; the landscapes appear on both grey and cream pages. Some pages have been carefully cut out, presumably because they contained unfinished sketches. One page with an unfinished watercolour of Everest from Chogorong has been cut out and stuck back in, evidently because it contains a finished landscape on the other side (Pls 80 and 107). A few extremely slight or insubstantial sketches have been omitted from this book, but some of the unfinished sketches, which are not without interest in their own right, have been included. For the rest, the sketchbooks are reproduced in their entirety.

All but a handful of the landscapes are identified and dated (sometimes on the facing page). The pencil annotations have become rubbed and faded over the years. So that the captions should not be lost, Norton's widow, Joyce, went over most of them in black biro. Occasionally, she misread the spelling or dates, as can be seen from a comparison with the entries in the diaries, which refer to the majority of the landscapes. Wherever possible, the captions to the plates reproduce those on the sketches, corrected by reference to the diaries where necessary. Many of the sketches of individuals and animals are, however, without captions; even when they do have captions, they are generally undated. It has been possible to identify many of the individuals, and a number of the undated sketches can be dated with more or less confidence from the expedition narratives. The drawings and watercolours are therefore reproduced in chronological order (the small number of undatable sketches being inserted where it seemed appropriate). Where captions have been made up (or added to), the editorial material is printed in italics. The letters (L) and (S) after the plate numbers indicate the larger or smaller size of the pages. The full-page landscapes from the larger sketchbooks are reproduced nearly full size, as are some of the pages from the small 1924 sketchbook.

The two expedition diaries are bound up into a single, leather-covered volume. The entries are written in a neat hand in pencil on ruled pages 6½in x 8in, generally only on the right-hand page of each opening. The text, fortunately, is still fresh and presents few problems of reading. The entries are consistent in format, and have remarkably few corrections, alterations or additions.

Place-names and personal names are printed as written. The spellings are not always consistent, and often differ from those printed in the expedition books. Where they differ significantly, mention is made in the Notes at the end.

Altitudes are given in feet, horizontal distances across the mountains in yards (e.g. 300x meaning 300yd). The summit of Everest is currently reckoned to stand at 29,035ft (8850m); in those days it was put at 29,002ft. Surveyors only reached Everest for the first time with the 1921 reconnaissance expedition, and Norton was well aware that even surveyed heights could not be considered precise, to say nothing of estimated heights.

A Glossary has been provided at the end, as well as a section of Notes. These explain or expand on particular points of detail in the texts which merit comment. A list of Norton's publications on Everest is included in Further Reading.

IMAGE CREDITS

Figs I, VI–VIII, X–XV, XVII and XX–XXIII are courtesy of the Royal Geographical Society. They are taken from the Photographic Library, Mount Everest Expedition 1922 and 1924 series, which contain photographs by a number of members of the expedition, including the official photographer John Noel.

Figs V, IX, XVI, XIX and XXIV are from the John Noel Photographic Collection, courtesy of Sandra Noel and the Royal Geographical Society. This includes some photos by other members of the expedition.

Fig. XVIII is by Howard Somervell, courtesy of David Somervell.

All other images are copyright Norton Everest Archive.

OPPOSITE: Fig. II   Map of the route from Darjeeling to Everest.

MAPS

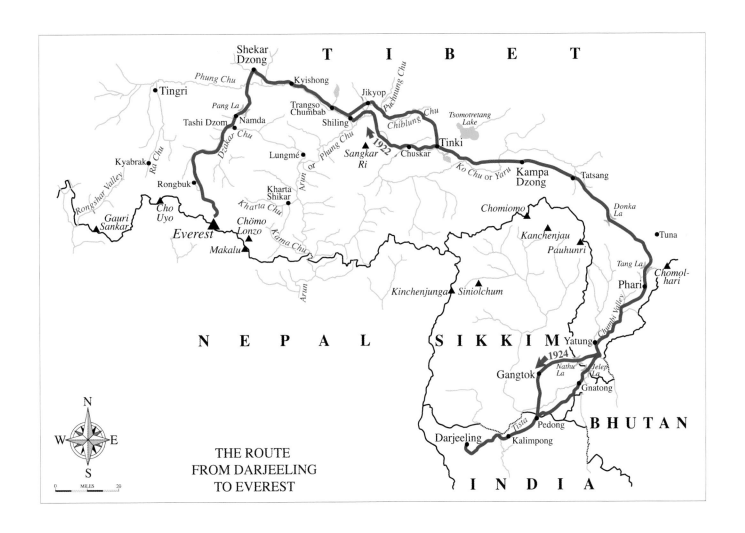

THE ROUTE
FROM DARJEELING
TO EVEREST

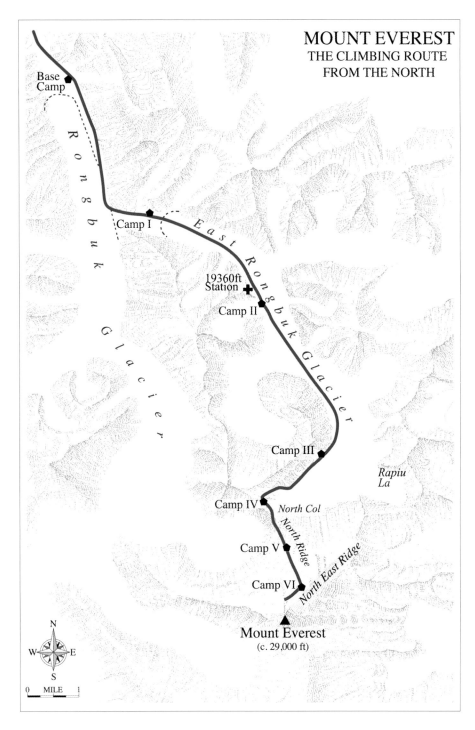

MOUNT EVEREST
THE CLIMBING ROUTE
FROM THE NORTH

Base Camp

Rongbuk Glacier

Camp I

East Rongbuk Glacier

19360ft Station

Camp II

Camp III

Rapiu La

Camp IV   *North Col*

North Ridge

Camp V

North East Ridge

Camp VI

▲ Mount Everest
(c. 29,000 ft)

N
W   E
S

0   MILE   1

LEFT: Fig. III   Mount Everest, the climbing route from the north. The 1922 expedition established Base Camp and Camps I–V; Camp VI was established by Norton and Somervell during the 1924 summit attempt.

OPPOSITE: Fig. IV   Map of the Everest region showing the rest period itineraries, to the east in 1922, to the west in 1924.

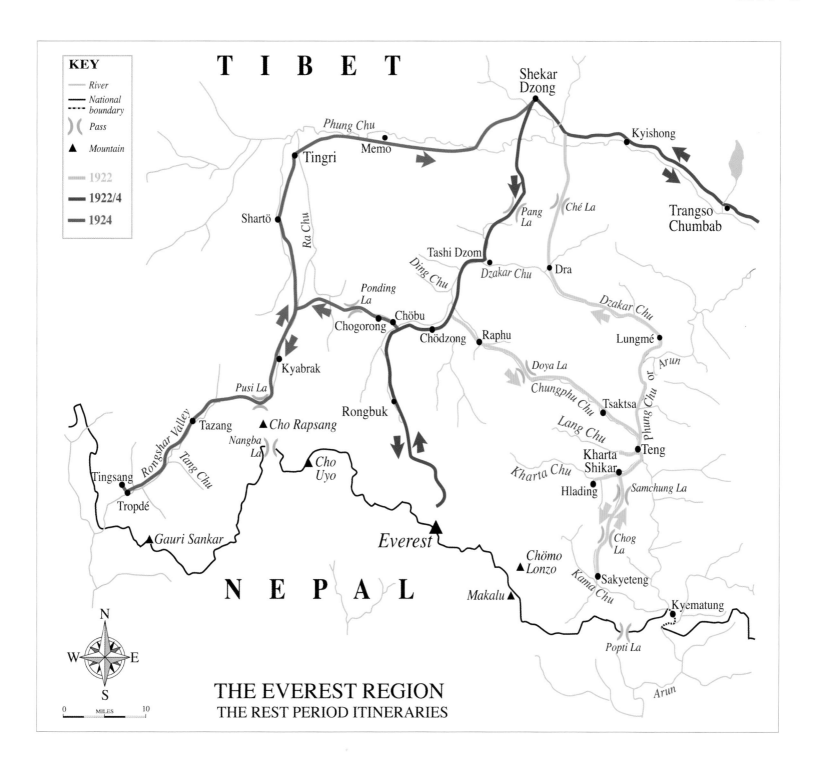

THE EVEREST REGION
THE REST PERIOD ITINERARIES

KEY

- River
- National boundary
- Pass)(
- ▲ Mountain
- 1922
- 1922/4
- 1924

TIBET

NEPAL

Shekar Dzong

Kyishong

Phung Chu

Memo

Tingri

Trangso Chumbab

Pang La

Ché La

Shartö

Ra Chu

Tashi Dzom

Dzakar Chu

Dra

Ding Chu

Dzakar Chu

Ponding La

Lungmé

Chöbu

Chogorong

Chödzong

Raphu

Kyabrak

Doya La

Chungphu Chu

Pusi La

Rongbuk

Tsaktsa

Lang Chu

Tazang

▲ Cho Rapsang

Rongshar Valley

Nangba La

Cho Uyo

Kharta Shikar

Teng

Tingsang

Tang Chu

Kharta Chu

Samchung La

Tropdé

Hlading

▲ Gauri Sankar

Chog La

Everest ▲

Chömo Lonzo ▲

Sakyeteng

Makalu ▲

Kama Chu

Kyematung

N
W E
S

MILES
0   10

Phung Chu or Arun

Popti La

Arun

Fig. V   E.F. 'Teddy' Norton. A portrait photo by John Noel taken at Everest Base Camp in 1924. The frostbite damage to the top of his ear sustained during the 1922 summit attempt is clearly visible.

INTRODUCTION

E.F. 'Teddy' Norton by Dick Norton and Bill Norton

While the diaries and sketches should largely speak for themselves, a brief account of Teddy's personality, career and interests may usefully serve to introduce them and throw light on the diary entries and his choice of subjects to paint. Born in 1884, Teddy's 70-year life spanned the reigns of six monarchs and two world wars. It started before the motor car, and at about the high point of the British Empire. It ended when two nuclear-armed superpowers were snarling at each other across the iron curtain, with the United Kingdom relegated to a supporting role. India and Pakistan were independent countries, while Tibet was under Chinese control. Teddy's attitudes and world view derived in large part from his upbringing in a late Victorian upper-middle-class family, but were very far from being limited to that stereotype.

A natural leader, Teddy chose the army as his career. He was a professional soldier for forty years, attaining the rank of lieutenant-general. The Everest episode came at the halfway point of his army service: he was a major at the time of the 1922 expedition, and a lieutenant-colonel two years later. He had been through the whole of the First World War, during which he was awarded the DSO and the MC, and was mentioned in despatches three times. No less significant, in the present context, was the fact that he had spent seven years in India before the war. He therefore brought to the Everest expeditions a familiarity with the country, its peoples and its languages. His knowledge of Hindustani and his facility for picking up the rudiments of other languages were to prove crucial assets on Everest, particularly at moments of crisis in 1924. Some of his subsequent career was also spent in India. Although best known for his achievements on Everest, he always regarded them, both at the time and in retrospect, as a sideshow to his main career. As an officer in the Royal Horse Artillery, he had daily contact with horses. In fact he was much more than a merely competent horseman. He enjoyed pig-sticking, and was for several years one of the leading competitors in the Kadir Cup, the prestigious annual pig-sticking event in India. Pig-sticking, while hardly in tune with present-day attitudes, was an uncontroversial and indeed exciting equestrian sport at the time. Teddy's love of horses is evident in the finely observed portraits of horses in the Everest sketchbooks.

For relaxation he enjoyed all field sports, especially (in his later years) fly-fishing. India provided opportunities for hunting game that could not be found in Europe, and the 1922 diary contains several references to stalking game during the expedition, including his dismay when a young companion abandoned the hunt, leaving two wounded animals in the field. He was frustrated in his desire to bag an ammon – the magnificent Tibetan wild sheep – but in 1924 such activities were not possible. Even if he had found the time away from his duties as expedition leader, the Tibetan government had forbidden any taking of wild creatures.

Teddy's climbing abilities were instilled in him in the course of many summer holidays at the Nortons' chalet near Sixt in the French Alps, which had been built for that purpose by his maternal grandfather, Alfred Wills, one of the eminent early Victorian mountaineers. Several of the diary entries and letters to his mother compare features in Tibet with those of places near Sixt. He also climbed more widely in the Alps and was a member of the Alpine Club. He had a great love of natural history. He delighted in seeing and recording sightings of mammals, birds, butterflies and wild flowers, and other natural phenomena. His familiarity with Alpine flora and fauna served him in good stead during the expeditions to Everest, and the diaries contain several entries comparing the Himalayan species he encountered with ones familiar to him from the Alps. Known for having this interest, he was appointed assistant naturalist to Dr Tom Longstaff on the 1922 expedition. This required him to shoot and skin birds for shipping to the Natural History Museum in London to enhance that institution's sketchy knowledge of Tibetan fauna, and to collect plant specimens for Kew Gardens. The 1922 diary reveals him devoting much time and effort to the collection of specimens ('Naturalist's job no sinecure' is one entry), an activity which must be understood in the context of an era when that requirement took precedence over today's focus on conservation. He kept separate notebooks on birds, and the diaries contain occasional detailed descriptions of particular sightings. He was also to have contributed a chapter to the expedition book on the flora observed and collected in 1922, but active service in the Dardanelles in 1923 prevented it.

Teddy was not normally a diarist, but his Everest expeditions were clearly special. So he kept the diary daily, apart from periods, such as during a high climb, when of necessity he wrote in retrospect, covering several days' activity in a single entry. The diaries paint a vivid, sparely worded picture of events as he saw them. They throw light on the culture of Tibet at the time, its landscape and its natural history, in addition to the mountaineering. His eye for landscape emerges clearly from the diaries as well as from the sketches. For instance, he described the landscape around Shekar Dzong in memorable phrases during both expeditions: 'lowish hills of the most wonderful variegated colours – pink, mauve, purple, scarlet & all shades of yellow and grey – all devoid of vegetation and bearing the dead burnt out appearance of yesterday's country' (24 April 1922); 'The colour on the variegated hills all around Shekar was wonderful – especially the crimson hill to the North – veined by deep blue cloud shadows' (23 April 1924) – no exaggeration of the astonishing Tibetan plateau scenery. His regular temperature recordings and other weather observations give the reader some idea of what the participants had to cope with throughout the expeditions. There are frequent mentions of the fearsome winds which buffeted them during the spring crossing of the Tibetan plateau, and Teddy's day-by-day account of the appalling weather and snow conditions encountered during the 1924 attempt on the mountain documents the toll they took of the climbers' strength.

He formed clear views about the personalities, strengths and weaknesses of other members, not to mention the Sherpas, in both expeditions. Indeed in 1924 it was part of his job, as leader, to do so. However, he confided few of his opinions on these matters to the diaries, mostly limiting references to his fellow climbers to matters of fact, rather than comment on personality or behaviour. Nonetheless, the strains of leadership during the moments of crisis on the mountain in 1924 are apparent, even if they are not dwelt upon. Nor are the diaries devoid of the lighter side of expedition life, though this is understandably confined in the main to the journeys out and the periods of rest and recuperation after the attempts on the summit.

Teddy was never a photographer and, unlike the other members of the expedition, did not take a camera. Instead, he took with him small bound sketchbooks and minimal painting equipment. There was little opportunity to paint on Everest itself, so there are only a handful of sketches of the climbers on the mountain. There are a number of sketches made around Base Camp and in the Rongbuk valley, but the majority are the fruits of spare moments during the approach and return marches and during the rest periods after the summit attempts. His subject matter is chiefly landscape, but people and animals feature as well. Indeed, Teddy had a special talent for sketches of this kind. His renderings of fellow expedition members and Tibetans alike are invariably full of character, and often witty. His accurately observed sketches of horses, donkeys, yaks, and wild animals such as the beautiful kyang (Tibetan wild ass) have real flair. There are also some notably attractive flower paintings.

Of the landscapes, one's eye is naturally drawn to his portraits of the mountains, among the most pleasing being those of Everest itself and Chomolhari, a mountain on the Tibet–Bhutan border which he much admired. No less interesting are his renderings of the Tibetan plateau, whose striking coloration he has so well captured, and of the lush Sikkim valleys where their journeys started and ended. Some unfinished sketches have a charm of their own.

All this work was executed with remarkable assurance and, doubtless, speed, in what must often have been very difficult painting conditions. Teddy mentions in his diary that his paint-water froze on one occasion. Those relentless Tibetan winds, and glare from the fierce sun, must also have posed real problems for the alfresco painter. The watercolour technique he mainly used was so-called 'body colour' on grey tinted paper, where white, or pale, passages such as snow or light cloud effects are achieved by opaque white paint. For the landscapes on cream paper he used the transparent watercolour technique, where clouds, etc. are formed by letting the ground colour show through. For those who are used to seeing the early Everest expeditions through the medium of black-and-white photography, the colour sketches of the landscapes are something of a revelation. Teddy's comment, in a letter of August 1924, was characteristically succinct: he compared Tibet without the colour to *Hamlet* without the prince. The sketches enable us to see the Tibetan landscapes through his eyes.

In conclusion, it might be thought remarkable that Teddy somehow found the time and energy to maintain his daily diary entries and to produce so many wonderful sketches throughout two serious mountaineering expeditions. The demands made by trekking and climbing at such altitudes would themselves be more than enough for many people. Yet in addition he was, in 1922, actively collecting animal and plant specimens, and for much of the time in 1924 fulfilling the heavy responsibilities of expedition leader, as well as personally achieving a world altitude record for climbing without supplementary oxygen that was to be unsurpassed for over fifty years. The material here needs to be appreciated in that light.

Everest 1922

The 1922 expedition was the first attempt to reach the summit of Everest. For decades the mountain had been the subject of distant fascination and frustrated aspiration among a small group of Himalayan experts and mountaineers, but no European had come anywhere near it. In 1921 a reconnaissance expedition was organised by the Mount Everest Committee, a joint venture of the Royal Geographical Society and the Alpine Club chaired by Sir Francis Younghusband. Its aim was to discover the approaches to Everest and assess the feasibility of an attempt on the summit. An approach from the south was not possible since the government of Nepal refused access, so the mountain had to be reached from the north, through Tibet. This necessitated a lengthy approach march of some four to five weeks, starting from Darjeeling in India. About 300 miles in length, the route had to be covered on foot or on horseback, with the equipment transported by mules or yaks. This approach imposed huge logistical problems, but it had one notable advantage: it enabled the expedition members to acclimatise slowly and steadily to the progressively thinner atmospheres at altitudes above 15,000ft.

The 1921 expedition succeeded in identifying and mapping the northern approaches to Everest. It also brought back valuable insights into the geology and natural history of the region, and much additional scientific information. It established that the best approach to the mountain lay past the Rongbuk Monastery directly to the north of Everest and that the North Col provided a feasible route to the summit. It also identified the most favourable period for climbing as being the latter part of May and the beginning of June, before the arrival of the monsoon. This would mean leaving Darjeeling before the end of March.

Encouraged by the success of the reconnaissance expedition, the Royal Geographical Society and the Alpine Club wasted no time in organising a follow-up expedition the next year. The members of the expedition duly gathered in Darjeeling in March 1922. Their leader was General Charles Bruce, a larger-than-life figure with unrivalled knowledge of the Himalayas and Himalayan peoples. His second-in-command, Colonel Bill Strutt, was the official climbing leader. The other members of the climbing party, apart from Norton, were: George Mallory, a leading climber of his day who was fascinated by the challenge of Everest, and who had climbed as far as the North Col during the 1921 reconnaissance expedition; Howard Somervell, a brilliant surgeon and a gifted amateur painter and photographer as well as an exceptional climber; Captain George Finch, an Australian with a particular reputation for climbing on snow and ice, and an enthusiastic

Fig. VI   Expedition members outside the Mount Everest Hotel, Darjeeling, March 1922. From left to right, seated: Strutt, General Bruce, Finch; standing: Crawford, Norton, Mallory, Somervell, Morshead, Wakefield.

advocate of the use of supplementary oxygen at high altitudes; Arthur Wakefield, another medical man and experienced climber; and Major Henry Morshead, a professional surveyor who had carried out pioneering survey work in the Everest region during the previous expedition. He was also an experienced climber with a reputation for toughness, which in the event was to be tested to the limit high on the mountain. The expedition doctor was Tom Longstaff, a fine climber and a Himalayan veteran, who doubled as the official naturalist to the expedition. He was greatly aided by Norton in collecting specimens and cataloguing the wildlife and flora. There were three transport officers: Colin 'Ferdie' Crawford, a former Gurkha officer and an Alpine Club member; and two Indian army officers, Captain John Morris and Captain Geoffrey 'Geoff' Bruce, a cousin of the General. Although a novice mountaineer, Bruce would impress with a very high altitude performance on the mountain. (He is referred

to in the diaries as G. Bruce, G.B. or Geoff, to distinguish him from his relative, who features as Gen(eral) Bruce, the General, or sometimes simply 'Genl'). The team was completed by the expedition photographer, John Noel, a remarkable adventurer with a long-standing interest in Tibet and a flair for mountain photography. His documentary film of the 1922 expedition was a remarkable achievement with primitive equipment in the most difficult of circumstances. Of the large contingent of porters, Sherpas and Gurkhas on the expedition, a number are mentioned by name in the diary.

Darjeeling to Everest

The diary begins on 27 March with Norton's first day's journey from Darjeeling to Kalimpong, partly by motor. After that, the expedition members proceeded on foot or horseback for the rest of the journey, accompanied by their large baggage train. At Kalimpong, they split into two groups, Norton travelling with the General in the first party. Following in the footsteps of the 1921 expedition, they followed a route across the south-east corner of Sikkim (Fig. II), through steep valleys clothed in tropical jungle. Norton delighted in the lush vegetation, the butterflies and the bird life. He helped Longstaff recording and collecting specimens, and began sketching the landscape. On 2 April they crossed the border into Tibet at the high pass of Jelap La, from which they had their first view of Chomolhari, a strikingly beautiful mountain on the border of Bhutan and Tibet that captivated Norton – he painted it several times. A long descent into the village of Yatung led them through much drier scenery that reminded Norton of his beloved Alpine haunts around Sixt; but for the first time the diary mentions the biting Tibetan wind. The march on towards Phari provided the first opportunity for hunting, when he and John Macdonald, the son of the British Trade Agent at Chumbi, set off in pursuit of burhel. Although not a full member of the expedition, Macdonald accompanied it for a considerable part of its journey, and is often mentioned in the diary.

At Phari the first party halted, to allow the second contingent to catch up, and prepare for the next stage of the journey. Here Norton was put in charge of the mess, and from then on had to devote much time to catering arrangements. The expedition then proceeded north-west from Phari to Tatsang, via the Tang La and the Donka La, heading towards Kampa Dzong. The four days before they reached Kampa Dzong were among the most difficult experienced at any point short of Everest itself. General Bruce later described 9 April as the hardest single march on the expedition. Norton's laconic diary entries are expanded on 11 April by a retrospect in which he sums up the relentless effect of the Tibetan wind which, with the cold, the snow and the long marches, had produced 'as severe conditions as we are likely to get later in the high mountains'. Two years later he confided to his diary that 'the Donka La is undoubtedly and invariably a stinker' (9 April 1924).

The expedition rested for three days at Kampa Dzong, from where they got their first distant view of Everest. They then continued westwards across the Tibetan plateau, through Tinki Dzong, Shiling, Trangso Chumbab and Kyishong. The diary for this phase of the march refers to the spectacular colouring of the barren Tibetan hillsides ('pink, mauve, purple, scarlet and all shades of yellow and grey'), the generally desolate terrain almost devoid of any vegetation, and the persistent biting winds ('like a toothache') blowing off the snow mountains to the south. The ten-day period was relatively uneventful, punctuated by some minor difficulties with the transport train. From time to time distant views of Everest appeared, and the party began to focus on the daunting challenge ahead. There were practice drills with the primitive and unreliable oxygen equipment, and some heated discussions about its value for the climb. After Kampa Dzong the diary shows an interest in night-time temperatures, and an increasing concern for fitness and evidence of acclimatisation. Occasional excursions were made to test climbing fitness on nearby cliffs or mountains. On 20 April, Norton's frustration at being unable to join the others in an attempt on the summit of Sangkar Ri shows through: 'I not for it.'

On 24 April they reached Shekar Dzong. The diary comments on the picturesqueness of the old fort and monastery on the limestone bluff of Shekar, but deplores the unwelcome attentions of the local people at their campsite. After two days at Shekar Dzong, the expedition turned southward for the final approach to Everest, the journey from here to Base Camp being described more fully in the first of Norton's letters home (see pp. 57–8). They crossed the pass of Pang La, which gave a magnificent view of Everest from almost due north, together with the whole main range of this part of the Himalaya. On Pang La they were struck by the extraordinarily small amount of snow on the north face of Everest. The descent from Pang La brought them, after a 'most delightful day', to a 'charming camp on a really green meadow at the junction of two streams … very peaceful and happy'. It is striking how often the diaries use the word 'happy' to describe the frame of mind of Norton and the others both before and after the assault on the mountain itself. Finally, a two days' march brought them to Rongbuk Monastery, 'in a wilderness of old moraine', 16 miles due north of Everest, to which it formed the gateway. In the afternoon the expedition was received at the monastery and blessed by the head Lama, a highly revered figure whose benediction did much to encourage the expedition's porters and the local labour force. The diary, however, makes no mention of this, recording instead the difficulties of sketching in the freezing conditions, and expressing increasing anxiety about the lack of news from home. The following day, 1 May, Base Camp was pitched 6 miles south of the monastery, below the snout of the main Rongbuk glacier, at an altitude of about 16,750ft. The yak and mule transport was paid off, and from here on all porterage would be by human muscle power alone. The assault on Everest itself was about to begin in earnest.

OPPOSITE: Fig. VII   Mallory (left) and Norton photographed by Somervell above Camp V during the summit attempt on 21 May 1922.

The attempt on Everest

The 1921 expedition had climbed as far as the North Col from the head of the East Rongbuk glacier. They had reached the head of the glacier from the east, by way of a difficult route over the Rapiu La, and had only belatedly realised that the East Rongbuk glacier itself provided the best means of approach. The 1922 expedition therefore had to pioneer a route from Base Camp the length of the East Rongbuk glacier (Fig. III). It took over a week to establish a path and to find suitable locations for Camp I (at the snout of the East Rongbuk glacier), Camp II (halfway up) and Camp III (at the head of the glacier at about 21,000ft). This became in effect an advanced base camp. Norton played a key role in this tricky exploration of the East Rongbuk glacier. His diary and the second of his three letters home provide a much fuller account of this vital step up the mountain than has previously been available. He returned to Base Camp on 9 May and, rather surprisingly, had his first hot bath since leaving Darjeeling!

After a four-day rest at Base Camp, Norton set off up the glacier again. Reaching Camp III on 15 May, he found that Mallory and Somervell had opened a route up the vast snow cliff above Camp III to the North Col. On 17 May he helped establish Camp IV on the North Col at about 23,000ft. This was his first climb to such an altitude, and caused him no difficulties. After a day's rest back at Camp III, Norton, Morshead, Mallory and Somervell began the first ever attempt on the summit of Everest. After a night at Camp IV, on 20 May they established Camp V at about 25,000ft, and passed the night higher than anyone had ever slept before. Next morning, Morshead was unable to continue, but the other three pressed on to a height of about 26,985ft, without oxygen. This was by some margin the highest altitude ever reached up till then; but they were forced to turn back about 2.15 p.m. Returning to Camp V, they found Morshead in a very bad way. Then began a nightmarish descent. A serious slip almost carried the four of them away; and Norton then had to support an almost incapacitated Morshead on his shoulder for hours down the ridge. Reaching Camp IV around 11 p.m., they were unable to

Fig. VIII   Norton photographed by Somervell leading towards the summit at the highest point of the climb on 21 May 1922.

heat anything to eat or drink. The descent to Camp III the next day took twice the expected time through fresh snow a foot deep. All four were frostbitten to some degree, but Morshead was badly affected. They staggered on down to Base Camp on the 23rd.

After this monumental effort, Norton remained at Base Camp recuperating for the best part of a fortnight, becoming increasingly sick of 'this beastly spot'. He sent home an account of the high climb (see pp. 62–6) and waited for news of the second attempt on the summit. Finch, an enthusiast for oxygen, accompanied by the novice mountaineer Geoff Bruce and Tejbir Bura, one of the Gurkhas, was forced to spend two nights at about 25,500ft at Camp VI by deteriorating weather. In spite of this, he and Bruce reached a height of about 27,300ft with oxygen, a new world record. They too returned to Base Camp frostbitten and exhausted. On 3 June, Mallory and Somervell moved up to Camp I again in the hope of making one final attempt on the summit. Two days later, unable to wait for news of the outcome, a party of semi-invalids, including Norton whose feet were still affected by frostbite, left Base Camp for lower and warmer parts. Finch and Morshead, accompanied by Longstaff, headed straight for Darjeeling, hoping to save Morshead's frostbitten fingers – some of which subsequently had to be amputated. Norton and the others turned eastwards for a long-planned rest period.

The rest period excursion and the return journey

The objective of Norton's party was the valley of the Arun river, one of a number of major river systems arising on the Tibetan plateau that flow southwards to break through the barrier of the Himalayan range into Nepal. They reached the Arun valley on 9 June at Teng, and descended a short way further to Kharta Shikar, the administrative centre of the district (Fig. IV). The diary comments vividly on the carpets of grass, spring flowers and flowering shrubs, and the abundant butterfly life, signalling the change of season since their first arrival at Everest – a constant theme of its account of the entire return journey. Norton kept busy collecting flowers and other specimens. With Longstaff no longer in the party, the job of expedition naturalist devolved to him. With hundreds of specimens to catalogue, he confided to the diary: 'naturalist's job no sinecure'. He was still suffering from frostbite on both feet and the top of one ear, and was periodically obliged to use pony transport instead of walking.

On 11 June a letter from General Bruce reached them by runner informing them of the outcome of the final attempt on the summit. On 7 June, Mallory, Somervell and Crawford, accompanied by thirteen porters, had set out to climb to the North Col. On the ascent they had been struck by an avalanche, and tragically seven of the porters had been swept over an ice cliff to their deaths. This news 'cast a gloom over everything'. The remaining members of the expedition rejoined Norton's party on 17 June at Teng. Two days later he organised a champagne picnic in a beautiful valley he had discovered south of Kharta Shikar. General Bruce later described the banquet in the expedition book as 'an epicurean feast', while Norton noted laconically in his diary: 'picnic quite a success'. The party spent six days at Sakyeteng in the valley of the Kama Chu, from where they glimpsed Chomo Lonzo, a major Himalayan peak east of Everest. Norton spent much of the time in the company of Mallory. They stayed a few days in a valley which they named the Valley of the Lakes, so full of flowers that Norton christened it 'my garden'. And then back to Teng. In spite of a certain amount of monsoon rainfall, the whole account in the diary is redolent of relaxation and delight in the spectacular scenery and spring growth.

On 5 July they finally began the march back to Darjeeling, by which time Norton reckoned that he was at last 'pretty sound now'. Travelling northward up the Arun valley, and then the Dzaka Chu, they followed a previously unexplored route over the Che La and thence back to Shekar Dzong, where Norton found the 'people as trying as ever'. From here the return march across the Tibetan plateau retraced the outward route, except for a diversion north of Shiling via the Chiblung Chu, through Jikyop, over a high pass and rejoining the former route at Tinki via its large and picturesque lake. The diary details the journey back via Kampa Dzong and Phari, describing among other things Norton's continuing activities as a naturalist and his frustration at failing to bag a burhel or the magnificent ammon sheep. The crossing of the Donka La was again accompanied by rough and unpleasant weather. On 27 July they crossed the Jelap La back into Sikkim. 'Goodbye to Tibet and very sad to leave it', is Norton's diary comment. And so back to Darjeeling on 3 August.

Fig. IX   The first climbing party at Base Camp after their record-breaking summit attempt, late May or early June 1922. From left to right: Morshead, Mallory, Somervell and Norton (with a bandage over his frostbitten ear).

THE 1922 DIARY AND SKETCHES

27 3/22   *Kalimpong Dak Bungalow*

Moreshead, Somervell, Crawford & self started 7.30 a.m. in motor to Takdah Cantonment, say 12 miles.

Thence on foot. Starting 8.45 descended some 4500 ft. to Tistar Bridge – 11.30. Spent ¾ hour here & at 12.15 started to climb to Kalimpong (about 4200 ft in 5 miles), a stiff pull up & pretty hot at bottom. I was carrying a fair sized sack.

Reached bungalow at about 3.15 & had magnificent meal.

The rest of party arrived about same time having come by train from Siliguri to Kalimpong Road & then ridden up.

Jolly bungalow but no view, threatening & looks like rain. Tried sketch – complete failure.

28 3/22   *Pedong (5000 about)*

All breakfasted with Dr Graham, a fine old Scotch missionary, the father of Kalimpong & founder of its wonderful schools & institutions. A.1. Scotch breakfast. Then on to a great meeting of boy scouts & girl guides to whom Gen Bruce read a message from Baden Powell.

Got away about 10.30 & the first party consisting of Genl, Longstaff, Mallory, Noel & G. Bruce & self marched to PEDONG 4300. Most rode. G. Bruce & I walked &, slinging along at a good pace, got in first arriving 13.10.

A long dull slog for about 9 miles on a gentle uphill gradient to a charming pass over forest-clad summit of ridge at Piengaon – & then descended 3 miles to Bungalow. Ugly cultivated country except crest of ridge where forest trees magnificent. Slack p.m.

Pl. 1 (L)   Rangpo Chu, Sikkim. 29 3/22.

29 3/22   *Rangli Chu (2700 ft)*

Marched about 8 a.m. & descended some 2300ft. to Rangpo Chu (say 2000), a lovely valley clothed in luxuriant tropical jungle. Tree ferns & magnificent trees. Butterflies of the most gorgeous varieties swarmed. Here we photographed, bathed, sketched [*Pl. 1*] & were very happy. Then climbed some 3000ft to Ari bungalow where we lunched & spent a couple of hours. On again at 2.30 & descended to the bottom of valley through magnificent forest scenery again. Pleasant bungalow.

Longstaff & I out after tea looking at birds. Saw Indian & white-capped redstart, the sombre dipper & the grey (?) wagtail among others.

30 3/22   *Sedongchen (7000 ft)*

Marched about 7.45 & climbed steadily & fairly steeply all along the sides of a long nullah, getting gradually into more & more open country (tho' there is plenty of big forest above us), air improving all the time. Walked with G.B., the others riding. We arrived about 11.30. In p.m. sketched [*Pl. 2*], stalked birds with L. & after dark skinned.

Before starting shot a redstart at Rangli & found it to be plumbeous redstart.

At Sendongchen saw among others striated laughing thrush, cinnamon-bellied nuthatch, lots of tits, green backed & others, including the beautiful crested tit I skinned.

The hills here covered with white violets & strawberries, both in fruit & flower (quite tasteless). Found a wonderful arum in wood.

Still too hazy for scenery to be good.

Cool night under 2 or 3 blankets.

31 3/22   *Gnatong (10500 or 11000 ft)*

Started 7.45 walking with G. Bruce – Gen Bruce also walked nearly all the way: rest rode.

The most wonderful march – road mounted directly from 7000 to 12000, first through magnificent forest full of magnolia in full bloom –

Pl. 2 (L) Sedongchen, Sikkhim, 6700ft. 30 3/22.

then came a stage of crimson rhododendrons in full flower – & some white ones – then big shrub rhododendrons (not yet in flower) & pine trees.

At Lingtu village had some excellent tea & chapattis provided by a cheerful Gurkhali lady. Crossing to the north side of Lingtu summit we got onto snow. Here the only flowers were the tiny pale blue gentian, a lovely mauve pink primula growing in dense beds & a little golden star.

Gnatong village is at the mouth of a regular Scotch glen much like a Cashmere stag valley.

G.B. & I reached Lingtu in 3 hours & the bungalow in about 4.

After lunch Longstaff & I out after birds. I sketched. Shot a wren; saw very few birds except large flocks of finches migrating – the only ones I saw close looked like Walton's twite.

I also saw some rose finches – probably the scarlet finch.

Snow still drifted thick against the bungalow – glorious wood fire.

This completes a wonderful 3 days. From Rangli Chu to Lingtu is a continuous climb of some 10000ft passing through every stage

Cho Traki
from just below
Jelep La

Pl. 3 (L)   Cho Traki from just below Jelep La. 2 4/22.

of vegetation & bird &c life from tree ferns & marvellous tropical butterflies to stunted pines & snow – a country almost devoid of animal life except migrants & a few crows & raptors.

No signs of game so far.

1 4/22

Late start 9.30. Short march 5 miles over two cols which opened up much more mountainous scenery including a considerable lake & the Jelap La – country bare of trees – or just the limit of tree line.

Got in 11.30 – soon came on to sleet & hail, so all spent a slack p.m. – skinning, writing &c.

Longstaff & I shot two blue-fronted redstarts & a stoat. Saw snow finches, choughs &c.

One or two of party not too fit, headaches & tummies – I so far A.1.

2 4/22   *Yatung Bungalow*

Started 6.30 with Mallory & G. Bruce to climb mountain overlooking Jelap La on N. Felt rotten & was sick half way to pass, so decided not to do mountain.

Slept a bit on far side of pass – wind bitterly cold – descended a bit & sketched [*Pl. 3*] – descended more & slept for a bit. Finally rode in last 5 or 6 miles on pony.

A very long march – quite 20 miles, descending from 14500 on Jelap La to 9800 at Yatung.

Scenery quite different – entirely conifers most of the way, firs (webbiana) high up, pines below. Hardly any green grass or signs of spring. The only flowers a primula exactly like Alpine farinosa – lots of these (denticulata), also a daphne very common.

Struck the main Chumbi valley about 5 miles below here – fine rushing river. Lots of white-faced wagtails & white-capped & plumbeous redstarts – also 2 ibis bills & any number of ordinary crows & choughs.

Villages exactly like Alpine villages – chalets identical in type.

On the Jelap was a good deal of fresh snow from last night. Fine view of Chumulhari to the East.

Got in 4.30p.m. – feeling better but tummy a bit wrong.

3 4/22   *Yatung*

Slept like a log 9 hours & woke completely recovered.

Shot birds with Longstaff from 11.30 until 4.30 up a delightful side nullah – (see list of birds). Great shikar after the ibis bill – wading stream to recover him. Shot a mouse hare. Busy skinning all evening.

After dinner were treated to a devil dance by the local inhabitants – quaint show.

2nd party arrived & spent night with us.

4 4/22   *Gautsa bungalow*

Longstaff & I off at 8 a.m. & shot all way to Gautsa, he riding, I walking. 12 mile march & a climb of 2000 or 3000 feet. Saw & shot a lot of interesting birds (see bird notes). Arrived about 3p.m. & spent rest of p.m. skinning hard.

The most delightful march yet, getting wilder & finer all the way with wonderful views of peaks (now snowy) looking back. Followed a boiling mountain torrent all the way. About halfway this ran for a mile through the bed of an old lake – level grass plain showing as yet no signs of spring green. Here we saw several great snipe & shot one.

After this scenery became rather like a glorified edition of Bérard valley – quite the finest valley I ever saw.

The forests were mostly excelsa pine low down, giving place to Webbiana – then very fine big birch trees & junipers mixed with them – much rhododendron scrub not yet in flower – in fact up here there is as yet no sign of spring & not a single flower.

Later the wild roses (a bit lower down) must be wonderful. Looks a fine game country, wish I could spend a week here.

Temperature & weather perfect all day.

Rest of 1st party arrived about ¾ hour after us.

5 4/22

Started at 5.20 a.m. with young Macdonald son of Trade Agent of Yatung to shoot burhel. Rode & walked 5 miles up valley to where it emerges on Thibetan Plateau. Here almost at once saw a herd of burhel on hill above. Stalked & alarmed them; climbed to ridge (nearly if not quite 16000ft). Saw same herd on far hill – very open hill side. I climbed right above them & made a most interesting stalk almost in full view all the way; got within 300x but couldn't get a yard closer – about 2 p.m. They began to graze, took long shot at best ram, underestimated distance & missed – Macdonald got shot at similar range & also missed. We both felt altitude a bit & went d–d slow. Missed ponies on way to Phari & had a longish walk, MacD. very beat; got into Phari 5.30 very hungry.

Saw a lot of interesting birds & beasts (see notes) – magnificent views of Kinchenjunga, Chomolhari, Cho traké & [*does not complete list*]

6 4/22

Worked at mess which I am to take over until 11.30 when MacD's shikarri came in & reported good ammon quite near. Rode out 1 hour but found he had been scared by local people. After searching a bit came home, arriving about 2.30 to find second party just in. Spent rest of p.m. arranging mess &c. Snow storm for about 2 hours. Hope to go for same ammon tomorrow if any khubber but have a lot to get through before we march day after tomorrow.

Views of Chomolhari & [*left blank for name to be inserted later*] in early light magnificent.

7 4/22

Looking at mess &c. in Phari all day bar a short walk shooting with Longstaff & Strutt in p.m.

Pl. 4 (L)   Chomolhari from near Phari. 8 4/22.

8 4/22

Marched some 20 miles to a desolate spot called Lung gye Dôk. Gorgeous views of Chumulhari at first – tried to sketch [*Pl. 4*]. About 11 a.m. came on to snow – & we marched until about 4.30 p.m. in a blizzard, & then camped in a most desolate spot. We crossed above i.e. W. of Tang La – & marched all day over absolute desert, only seeing one party of yak drivers in a miserable little tent.

Longstaff's pony threw me off while I was trying to arrange comforter, & dragged me some 50x by stirrup, kicking me.

Got off without a scratch.

Got a meal of sorts about 5.30 & all to bed. Slept 11 hours like a log. Several of party knocked out by cold, sick.

9 4/22

Brilliant morning – about an inch of snow over everything. Got off at 8.45 & marched till about 5 p.m., over 20 miles – probably 24, crossing two passes over 17000.

Rode a bit. Bitter wind off the snow mountains in p.m. Like a toothache, couldn't get away from it. Saw kyang.

All pretty weary. High tea 5.30 & early to bed, 3 or 4 in each tent. Camp under a low limestone cliff, place called Angzang Trag. Slept A.1.

10 4/22   *Angzang Trag (about 16500)*

Decided spend day here – 3 coolies still lost. Transpired they had fetched up at a monastery near here.

Brilliant morning – view & colouring simply magnificent. The three great massifs of Pau Hunri, Chomiomo & Kanchenjau showing up over western horizon quite close.

Spent day squaring up & skinning birds shot at Phari. Sun all day but bitter wind – boots & everything frozen stiff in morning.

This is a completely desolate gravelley plain with rounded pink & yellow hills cropping up. Saw some interesting birds.

11 4/22   *Kampa Dzong (15200)*

Marched at 8.15 – steady climb over wide gravelley plains intersected by occasional nullahs to pass about 17000. Two or three successive ridges & then descended to great plain on edge of which stands Kampa dzong. Most picturesquely situated under a bluff on which are the Dzong & Kellas' grave.

After descending from pass began to meet vegetation – dry scrubby grass & very low scrub. Pleasant change, as all day & for last 3 days have seen nothing in shape of vegetation except rounded lumps of some plant growing like a sponge & looking like the rounded tops of boulders – all dead now & varying russet colours. A little dead grass at last camp along stream.

On far side of pass plain was covered with herds of kyang (at least 100 in sight at a time constantly) & gazelle.

Thought we saw sheep (burhel or ammon) on range of hills which parallelled our course all the way but too far off to swear to.

The colours both on the great extent of pink & yellow plain (rather like Sinai Peninsula) & in the great extent of blue sky flecked with clouds were wonderful. Great snow mountains to S. & S.W. all day – lots of cloud but sun all day.

The whole southern skyline of Kampa dzong plain is bordered by snowy range forming N. boundary of Sikhim.

Minimum thermometer registered 8°F. last night which was very warm compared with night before, when stream was frozen stiff & thermometer must have registered zero.

I slept v. badly.

It is hardly necessary to note the wind – that well known feature of Tibet. Morning usually breaks still & sun is so hot one can strip & wash in frozen stream & have breakfast in the open. By about 9 the wind (generally S.W.) is blowing bitterly keen off the snow mountains & increases in force until it begins to die down again at sunset when night is generally still. This wind, as I said before, is like a toothache – one can't forget it. Coupled with the low temperatures at night, the blizzard on 8th & the long marches it produced perhaps as severe conditions as we are likely to get later in the high mountains. So far I have stuck it well. God bless Farrar & his 7ft. sleeping bag.

We all go pretty badly at 16000 or 17000 up hill so far. One has to breathe deliberately in fancy ways & one's pace is lamentably slow.

12,13,14 4/22

Spent at Kampa dzong – welcome slack – did nothing much but mess about in camp & stroll a mile at a time sketching or looking at birds. Finch & Crawford turned up on 13th looking the most awful ruffians (as they seemed to think us). They had come through without a halt encountering a good deal of snow which was a foot deep in places.

13th & 14th were gorgeous days – hot sun & hardly any wind. Yet sketching in blazing sun on 13th at 10 a.m. my water froze stiff on paper in a moment when I held paper in my own shadow [*Pl. 5*]. 19° to 24° of frost each night.

Fine view of Everest on 13th.

Spent a good deal of time overhauling mess.

Pl. 5 (L)   Mt Everest from Kampa Dzong. 13 4/22.

15 4/22   Linga (about 15000)

Marched 8 a.m. 17 or 18 miles to Linga – dull march over dead flat sandy or gravelley plain rising over a low hill for middle 5 miles which I rode.

Just before this crossed Ko Chu or Yaru river – a fordable stream.

Sun very hot. Gorgeous day, wind only bad towards 3 p.m.

On arrival Longstaff & I out looking at birds – tufted duck, brahminy & bar-headed geese – the latter so tame we photographed them at 30x or 40x.

All these on a big ghil round village, mostly dry.

Gorgeous views of mountains along Sikhim boundary – now quite close on the south.

16 4/22   Tinki

Started 7.45 & marched some 14 miles. I rode all way for first time. Dead flat march over semi dried marsh largely salt – then over sand dunes & then over a succession of low rocky spurs separated by flat valleys like polo grounds. On our right curious rounded red hills all the way – on the left the Sikhim boundary snow mountains all the way. Lovely views & colouring all day – temperature delightful – glorious weather – wind only became bad about 3 p.m.

Villages (there are 2 or 3) situated on a shallow lake which is full of duck & geese.

In p.m. L & I took a stroll after birds: I shot a finch we can't identify with catapult. After tea tried to sketch but water froze as usual [Pl. 6].

At and after dinner tremendous argument oxygen versus no oxygen. Delightful day.

17 4/22

Spent day at Tinki dzong – Longstaff sick.

I out with telescope identifying duck &c – then skinned finch until lunch. In p.m. a rock scramble with Mallory, Somervell & Crawford; after tea oxygen drill.

Decided remain here tomorrow on account of Longstaff.

18 4/22

Remained Tinki on account of Longstaff who is better. Oxygen drill, mess accounts – then to lake with Finch & telescope after birds. In p.m. sketched up above Dzong. Very cold & unsettled weather, wind bitter.

19 4/22   Chuskar

Marched 7.30 over Tinki La, 17000, to Chuskar. I walked all way bar last 3 miles on flat – a stiff climb at end. Not yet at all satisfied with my wind over 16000.

Lovely day on Tinki side of pass. I climbed without coat & sweater. On Chuskar side met bitter wind which increased to an unbearable gale in valley. I arrived in camp feeling quite bad; laid up under a wall in sun for 2 hours in p.m. & was quite O.K. by tea.

This wind beggars description – & ruins this country almost every p.m.

20 4/22

Lovely morning disclosing magnificent snow peaks of Gyangka range at S. end of valley – last night obscured by clouds & dust – one striking peak with fine precipices.

Marched 8.30 after some trouble with transport. (1st time).

After about 6 or 8 miles met Yaru or Ko Chu River again (Mendé last time) & forded it, spending an hour or more on bank to get our kits across dry. Pleasant interlude with much photography.

Some 2 miles on at Gyangka Nangpa found our advance transport halted & camp pitched – some hitch over transport – so instead of going on to Rong Kong remained here.

I rode most of this march.

In p.m. Mallory, Finch, Wakefield & Somervell started to climb Sangkar Ri, the mountain above here – intending to camp high tonight – come back here tomorrow after reaching summit & then do 20 mile march on ponies. I not for it.

L & I spent an hour or so after birds seeing a linnet we can't identify. Wind not so bad this p.m.

21 4/22   *Shiling (junction of Yaru & Arun)*

Some delay over transport. Marched about 10 all down valley of Yaru – a very desolate stony valley, the only vegetation being Tibetan Gorse now all bare & brown (a very few flowers seen just like our gorse). Lovely day, little wind & quite hot at times. Rode a good deal.

Finally forded stream & camped under a fine crag of limestone & slate (?) overlooking a jolly sheltered valley.

After 3.30 tea Crawford & I climbed face of crag – 1hr 20 min to top (about 1500 feet) & 10 minutes down – (a splendid sliding scree).

Finch & Wakefield fetched up by 3, Mallory & Somervell by 8 p.m. None reached top tho' latter pair got near it.

22 4/22   *Trangso Chumbab*

Marched about 8. Emerging from gorge were met with magnificent view of Everest blocking end of valley & reflected in the still waters of the Yaru. Most striking picture – all photoed or sketched [*Pls 7–8*].

Then crossed subsidiary stream & some miles of sand dunes & quicksand. Turning into main Arun valley marched 5 or 6 miles up pleasant meadows between river & red sandstone hills – these meadows <u>almost</u> green in places & very pleasant going.

Turned away from R. up tributary valley & camped in a pleasant open meadow between sand or rather gravel dunes. Lovely day – quite warm & wind nothing much.

Pl. 6 (L)   Tinki. 16 4/22 [*perhaps should read 18 4/22*]. *The peak on the left is labelled* Chomiomo.

Pl. 7 (L)   *Pencil sketch similar to Pl. 8 of Everest from near Shiling, presumably 22 4/22.*

Pl. 8 (L)   Mt Everest reflected in R. Yaru near Shiling. 22 4/22.

23 4/22   *Kyishong only 8° of frost in the night!*   **24 4/22**   *Shekar*

Off at 7.45 – a longish march of some 18 miles. I walked & rode about half each. For 1st 7 miles we crossed a level gravelley plain – the surface of one of the old river terraces overlooking the river.

Then descended to a lower terrace & then another to present river level. Very surprised to find regular forest of sea buckthorn along river in places all quite bare now of course – but first trees we have seen in any numbers.

From here on marched over pleasant meadows (almost green) for some miles, enjoying delightful views up & down valley.

Further came into most desolate country we have yet seen – hills much like those on the moon, absolutely devoid of vegetation & looking rather like heaps of burnt out red cinders.

Got into a grubby village camp about 2.45 – after some grub L. & I out after birds – a pleasant & interesting stroll.

Picked a tiny gentian & new pink primula – saw choughs collecting nesting material. The Tibetan spring is really beginning.

Only 11° of frost in the night.

Marched 7.30 along gorges of the Phung Chu until we emerged after about 2 hours onto a wide open plain representing the junction with Phung Chu with Gyal Chu on which Shekar stands. Turned up this valley.

This plain the usual gravelley wilderness but surrounded by lowish hills of the most wonderful variegated colours – pink, mauve, purple, scarlet & all shades of yellow & grey – all devoid of vegetation & bearing the dead burnt out appearance of yesterday's country.

Lunched by a most pleasant stream with some almost green turf. The big bridge over this broken & never repaired as usual. It seems as if since the Chinese ceased to run this country the local inhabitants have let everything go to pieces.

Everywhere the number of ruins indicate a decreasing population. The country is evidently in process of drying up & no doubt supports less than it did.

Shekar most picturesque. The old fort & monastery respectively on top of & half way up a steep isolated pinnacle of limestone – the village on the flat below.

Got settled in by 1 or 1.30.

Weather both yesterday & today quite perfect – only enough wind (so far) to counteract the considerable heat of the sun.

25 & 26 4/22

Remained at Shekar. Filthy camp with driving dust storms. Tibetans a great nuisance – utterly devoid of manners & treating us like a wild beast show, crowding into camp & stinking the place out.

Sketched [*Pl. 9*] – climbed Dzong (1000ft above camp), wrote letters & sorted mess stores.

Shall be glad to leave here.

Pl. 9 (L)   Shikar Dzong. 24 4/22 [*perhaps should read 25 4/22 or 26 4/22*].

Pl. 10 (L)   Pangla 27 4/22.

27 4/22   *Pangla*

Marched 8.30 & walked all way – only 9 or 10 miles. First through lifeless 'lunar' hills of all shades of yellow, pink & brown. Then descended to Phung Chu & spent an hour or more loafing by the bridge – a solid affair of 4 or 5 piers. Found majority of transport had gone another way & forded river so turned W. up S. bank & joined them about 1.30 in a delightful grassy camp some ½ mile from a very small village. Delightful change from that foul spot Shekar. Basked in sun, washed in stream & were all very peaceful & happy.

Very pleasant temperature all day: walked comfortably in shirt & tweed coat, sweater over shoulders, & yet was warm in camp without a tent.

28 4/22   *Namda*

Marched at 7.30 – walked all way, some 10 or 11 miles over Pangla (about 17000) & down to camp, about 14500.

Walked A.1. today: immense improvement on Tinki La day. Am undoubtedly acclimatizing. Lot of ram Chukhor near top.

A most delightful day. Made pass in 1 hr. 45 min. going very easy. Grand view of whole main range from top. Sat & sketched for an hour or more [*Pl. 11*] – Wakefield found delightful gentian (like small accaulis) on absolutely bare sandstone scree, also saxifrage. I caught small apollo-like butterfly – Longstaff a lizard.

Descended through fine barren sandstone gorge (where G. Bruce, preceeding us, saw herd of burhel – easy shot for buck!) – the first game since Kampa).

Saw Godlevski's meadow bunting.

Charming camp on really green meadow at junction of 2 streams – away from village. Very peaceful & happy.

On pass there was a tremendous argument as to whether great rock peak facing us was Everest or not – the reason being that whole N. face appears extraordinarily devoid of snow – in striking contrast to E. face seen from Shiling.

Ultimately no doubt about it.

29 4/22   *Chödzong*

Very cold morning: 20° of frost at night again. Marched 7.30 – 2 miles to valley of Dzakar Chu – turned up this for some 14 miles. An ugly valley – flanked by brown rounded hills. Weather inclined to look threatening – sun often obscured & beastly cold wind.

Walked most of way – & arrived in camp about 1 p.m. Slack p.m. – pretty cold. Now snowing lightly (8 p.m.).

30 4/22   *Rongbuk*

Bitter cold morning – back to every bit of warm clothing. Marched 7.30 – soon left every vestige of herbage except a few juniper bushes & cistus. Marched for miles up a barren wilderness of rock – the Rongbuk stream an absurdly small affair (about size of Guivra stream). Warmed up by fits & starts as one got shelter from wind or sun came out. But unsettled

CHO UYO (R) 26870ft
GACHUNG KANG (L)
from PANG LA
28 4/22

Pl. 11 (L)   Cho Uyo (R) 26,870ft, Gyachung Kang (L), from Pang La. 28 4/22.

all day with periodic snow flurries. Made Rongbuk about 1 p.m. A big monastery in a wilderness of old moraine.

V. cold p.m. with occasional gleams during which almost warm.

Most magnificent view of Everest – mostly only glimpses but all clear for a few minutes at sunset. Sketched, but everything froze – a pity – best view yet [*Pls 12–13*].

Getting v. anxious for mail & some news from home.

1 5/22   ~~16500~~ *17000ft Camp*

Marched some 6 miles to near foot of Rongbuk glacier where all transport struck, about 1 p.m.

Pitched a permanent base camp in a basin between moraine heaps, & spent p.m. settling in & selecting mess stores for higher camps.

Weather mending but still very cold the moment sun goes in.

Tuppoo (my bearer) left sick at Rombuk monastery.

2 5/22   ~~16500~~ *17000ft Camp*

Lovely morning with delightful hot sun – wrote home mail.

At 10.20 Strutt, Finch & I went off to reconnoitre No. 1. camp. Following old moraine terrace some 1500 ft. above bed of valley, we went some 5 miles round corner in E. Rongbuk valley & finally selected camp near snout of that glacier at about 18000.

Had lunch there & returned in a little under 2 hours following the trough between R. Bank of Rongbuk glacier & moraine.

Overcast p.m. & very cold again.

18° to 23° of frost the last two or three nights.

3 5/22   *17000ft Camp*

Glorious hot morning after 23° of frost at night – clouded over in p.m.

Spent slack day in camp, helping sort tents &c in a.m., sketching in p.m. [*Pl. 14*].

4 5/22   *17000ft Camp*

Spent whole day sorting tents & mess stores for various camps. Came on to snow in p.m. – whole valley white. Mallory & Somervell climbed 21000ft peak on opposite side of valley.

5 5/22   *17000ft Camp*

Lovely morning. Snow soon gone.

Reconnaissance party (Strutt, Longstaff, Morshead & self) with 16 coolies started about 10 a.m. for No. 1. Camp with a view to selecting Nos 2 & 3 camps.

Plugged up trough on E. side of Rongbuk glacier & then up frozen E. Rongbuk stream.

This now melting fast – last time it could be used as a highway.

Arrived at No. 1. Morshead & I went on in p.m. halfway to No. 2. & back to evening meal at about 5.30.

All turned in in sangar with Whymper fly roof & spent a moderate night.

Pl. 12 (L) Pencil sketch of Everest from a viewpoint similar to Pl. 13. ? 30 4/22.

Pl. 13 (L)   Everest from Rongbuk. 30 4/22. Everything froze.

17000ft
Camp 3 5/22

Pl. 14 (L)   17,000ft Camp [*i.e. Base Camp*]. 3 5/22.

6 5/22   ~~No I~~ No II Camp (about ~~18000~~ 19000)

Started about 8 a.m., M. & I ahead. Got on O.K. as far as tributary glacier just beyond Wheeler's 19,360 photographic station. Glacier stumped us at first so camped near W.'s station. Strutt & Morshead found way across glacier some way up. M. & I then found better route & suitable spot for No. 2 camp by crossing snout of glacier on moraine of E. Rongbuk glacier.

Returned for evening meal at about 5.30 p.m. This from now on consisted of a 'hoosh' compounded of one or two A. & N. rations, one or two tins of soup, crushed biscuit & coolie's 'sampha'. Breakfast (also cooked on primus stove) of tinned sliced bacon & sardines & tea.

2 Primuses going – coolies cooking similarly. Longstaff & Morshead superintending. 'Poo' soon learnt to carry on.

Since leaving No. 1. Camp not a vestige of vegetation or a living thing – only glaciers & moraines.

Our route all day up latter – vile walking.

Slept A.1. – most comfortable in eider down flea bag – without blankets – 2 in an 'improved Meade Tent'. Thermometer down to a few degrees above zero.

7 5/22

Moved camp to site selected last night – & there left Longstaff (who is none too fit) & coolies, of whom half were sent down.

Strutt, Morshead & I pushed on & found crossing over next (& much larger tributary glacier) principally by S.'s good glacier craft.

Weather overcast & windless. All suffered from horrible 'glacier lassitude' making one feel as if one hadn't a bone in one's body.

Encountered a lot of difficulty – all snow slopes turned out to be the hardest glassy ice & after struggling on to near corner of spurs of N. peak (below 20590) were defeated by a small but very 'serac'd' glacier & turned back.

Decided to sleep again at No. 2. & try a route up main E. Rongbuk next glacier day.

On our return route Morshead discovered easy crossing over great central moraine trough – the principal obstacle.

8 5/22   *No. 2 Camp (19000)*

Another good night tho' thermometer 2° below zero. Decided take all spare tents &c. on to No. 3. camp. Longstaff ill – apparently 'flu – so left him & 2 coolies in camp & went on with 5. Got across moraine easily & so onto broad open surface of glacier – which proved to be mostly hard nobbly ice.

Plugged up this to opposite our corner of yesterday when we encountered good snow surface slightly crevassed – so put on rope.

Lovely fresh air, enough but not too much wind, powerful sun (yet I wear only Cashmere puttoo hat).

Just as we were thinking of making best of a bad camp on corner discovered that by turning corner towards N. Col we found our ideal spot about on 21000 ft contour under cliffs of N. peak facing S.

Big open space of moraine & scree free of snow – a sun trap, sheltered from wind & with best water we have yet drunk in Tibet.

Here we lunched in great content – dumped loads (1 cooly – Dawa – by the way, had fallen out en route & his load been distributed among others) & spent an hour or so.

4 ½ hours from No. 2. camp. Well satisfied with our selection of No. 3. camp returned to No. 2. in 2 hours by same route.

Found L. pretty bad.

Thought it best for Strutt not to sleep in same tent so after tossing up I turned out into stone sangar roofed with tent fly – M. & S. sharing other tent.

Thermometer fell to 5° below zero in night – yet I slept badly from being too warm in eider down flea bag & my clothes.

9 5/22   *No. 2 Camp*

Longstaff insisted on struggling down to No. 1.

The rest of us packed up, dumping all available kit, & started down at 8.45 – glorious weather.

Made No. 1 camp about 11.45 – building cairns all the way. Lunched there – left Longstaff with Morris who arrived with coolies – & made base camp in 1 ½ hours.

Met English mail en route – at last. Sad news of poor old Uncle Jack's death & Dad's seizure.

Had a <u>hot bath</u> – first since Darjeeling.

10 5/22   *Base Camp*

Slacked & wrote letters all day.

11 5/22   *Base Camp*

Seedy – tummy – all day.

12 5/22   *Base Camp*

Slack day – recovering – bitter wind – 24° frost at night.

13 5/22   *Base Camp*

27° of frost. Wind not so bad. In a.m. walked up shelf ½ way to No. 1 & sketched [*Pl. 15*] – slack p.m.

14 5/22   *No. 2 Camp*

Strutt, Morshead & self started for No. 3 Camp & a definite attempt on the mountain.

Left 9.25	arrived No.1.	12.05.
	left "	13.25.
	arrived No.2.	16.00.

Fine day, no incident – turned in soon after sun left camp. Found Crawford here pretty mountain sick.

Pl. 15 (L)   Everest from Base Camp. *Probably 13 5/22.*

May 15ᵗʰ   *No. 3 Camp*

Left No. 2. 08.10
arrived No. 3. 12.05.

Went up with a caravan of 36 coolies: found glacier much changed – all surface snow gone at lower portion & very slippery ice surface.

Found Mallory & Somervell had established a route up to N. Col with ice in only 3 or 4 places & had fixed ropes at most of these.

Spent a slack p.m. Sun leaves this camp by 3.20.

May 16ᵗʰ   *No. 3 Camp*

Spent a slack morning: discussing plans &c.

In p.m. Mallory, Morshead, Somervell & self to Col at head of glacier overlooking Kama Valley (Rapiu La).

Bitterly cold wind – Kama valley a boil of clouds with impressive glimpses of Makalu – looked very like beginning of monsoon conditions.

For some unknown reason I walked entirely without difficulty this day.

May 17ᵗʰ   *No. 3 Camp*

All to N. Col to establish a camp & fix an extra rope.

Left No.3.	09.15.
N. Col.	13.45.
Left "	14.20.
arr. No. 3.	16.00.

No difficulty encountered – fixed ropes & good steps getting over any difficulties. Strutt was almost beat & only just made the Col.

Pitched tents on snow under shelter of big cornice or ice cliff.

May 18ᵗʰ   *No. 3 Camp*

Slack day with small trial trip to try coolies' new boots.

May 19ᵗʰ   *N. Col (23000)*

Strutt having decided not to take part, Morshead, Mallory, Somervell & self started about 8 a.m. for N. Col taking 10 coolies with a few loads to complete N. Col camp + 4 very light loads for proposed camp about 25000.

Reached N. Col early & spent p.m. fixing another rope for next day's advance, pitching camp, cooking &c.

From here on all cooking done with Meta or alcohol in spirit stoves – & a pretty miserable job.

May 20ᵗʰ   *25000ft Camp*

Mallory got us out at 5 a.m. only to find half the coolies mountain sick. Probably because all slept with tents tight shut. Got breakfast cooked & off by 7.30 or so with only 4 coolies (we had intended to have one spare cooly per load).

Morshead led most of way – went right up N. arête on scree just ~~E~~ W of snow. Wind very cold.

Lost my rucksack with all spare warm clothes over kad. About noon wind drove us off arête & we took shelter on ~~W.~~ E. side & decided to camp.

Hopeless place as no level ground even for our 2 Mummery Tents. Eventually got them up on bad sloping uneven rocky platforms & sent coolies (who by now were complaining bitterly of cold) back. Intended them to remain at N. Col but they all descended right away to No. 3.

Spent a very cold & miserable p.m. cooking under very adverse circumstances.

Turned in early & spent a pretty bad night – sleeping 2 in a bag – I found my right ear had been frost bitten & I couldn't lie on that side – which didn't help much.

May 21st   N. Col

Dawn found it snowing with an inch or two on the ground. However decided to go on.

After usual preliminaries got off about 8 a.m. Morshead chucked it at once as he found he was unfit to go – Mallory felt pretty bad: I led alternating with Somervell.

Going would have been very easy but for new snow which made it none too good.

We continued to follow arête until about 2.15 we found ourselves some 400 ft below big gendarme of N.E. shoulder – aneroid marking 26,750. Here we decided (wisely as event proved) to turn back.

Mallory now led: we descended to 25000ft camp, & picked up Morshead.

It soon became evident that he had 'crocked up'. He slipped on ice slope & became very slow – in fact could hardly get along at all.

I then slipped badly on snow slope & carried away everyone except Mallory who held well.

I now had to support Morshead on my shoulder & did so for some hours.

We made N. Col soon after dark – lit lantern & proceeded to grope for the rather complicated route through cornices & crevasses to our tents. ~~The morning's~~ Y'day's tracks were obliterated by about a foot of new snow. To make a long story short we got to camp just before 11 p.m.

Found food but no spirit stoves to melt snow (These were in coolies' tent all the time but we missed them & thought coolies must have taken them down to No. 3) & as we were all famished for thirst we ate little or nothing.

Slept A.1.

22nd May   No. 3 Camp

On waking decided not to breakfast without drink but to descend at once to No. 3 – which normally took 1 ½ hours or so.

On starting soon found we were in for a bad time – all tracks, steps & some fixed ropes were completely obliterated by a foot of new snow.

I led followed by Morshead, Somervell & Mallory.

Every step had to be kicked in a foot of snow with nothing really to keep it from sliding off – the very greatest care was necessary & it took a most exhausting 3 hours to reach the bottom – which we did without mishap.

Here we met Finch, G. Bruce, Wakefield & some 20 coolies with oxygen apparatus.

They most generously fed us on tea & brandy from Thermos flasks. As we were famished for thirst it is needless to say how we appreciated this.

Wakefield accompanying us we went on into No. 3 Camp where all – especially Strutt – looked after us in the kindest conceivable way – changing our boots while we fed, lending us their sleeping bags &c.

It was here found that Morshead had all fingers of both hands frostbitten & Mallory most of them.

Somervell one finger slightly bitten, in addition to my ear.

Rested all p.m. & slept like a log.

23rd May   Base Camp

Decided to try & get right through to Base Camp – on account of frost bites.

Started about 8.30 a.m. & got there by 5 p.m. lunching at No. 2 & tea at No. 1.

A wearisome trek during parts of which I could hardly stagger.

Morshead had trouble with his foot & found on arrival that one toe was frostbitten & had been overlooked yesterday.

Great reception by Bruce & Longstaff – whisky in tea – fizz & quails for dinner – then such a sleep.

24ᵗʰ May   *Base Camp*

Did absolutely nothing all day – complete worm.

25th May   *Base Camp*

Wrote home all day – still a worm.

26ᵗʰ May   *Base Camp*

Bad night – ear painful – wrote up diary – getting better. MacDonald arrived.

27–31 5/22

Practically lived in Base Camp recuperating – & not fittening up as quick as I should like.

Went short walks most mornings & did some ~~triangulating~~ theodolite work others [*sic*]. Weather consistently cold. Night min. temperatures increased to say 22°F or thereabouts but days were marked without exception by bitter blistering wind often starting at 8 a.m. & always continuing until 7 p.m. or so. Plenty of sun but one wore every sort of thick clothes all day long.

Everybody heartily detests this beastly spot.

Finch & G. Bruce returned from their oxygen attempt on 29ᵗʰ? – latter with both feet frostbitten – former completely exhausted. They had reached a point perhaps a couple of 100ft higher than our party & had had a very rough time.

1 & 2 6/22

Nothing to record but bitter winds & snowstorms up valley, miserable conditions in camp.

Have spent ~~2~~ 3 days trying fix by theodolite heights of various parties on Everest – but mountain obscured by clouds.

5 6/22

After 2 miserable days during one of which it snowed & we played poker all day – (Finch returned from Camp I unfit to go on) – marched on p.m. of 5ᵗʰ to Rongbuk.

6 6/22

Marched to Chodzong. Glorious day & enjoyed every moment of it. Flowers, birds, butterflies & green grass once more. Country much come on since we were last here. Everything a growin & a blowin (for Tibet).

[*left hand page*]
N.B. Longstaff left with Finch & Morshead for Darjeeling. It was imperative to get last to a milder climate quickly to save his hands.

Geoff Bruce, MacDonald & I left same day & both parties camped together at Rongbuk Monastery.

Next day we separated, they heading for Shekar & home, we for the Kharta valley as advd guard to the remainder, who were to follow us after another attempt on the mountain by Mallory & Somervell.

7 6/22   *Raphu (15300)*

Marched at 8.45 & did first 4 miles with Longstaff & party. Lovely morning, warm & hardly any wind. Lots of birds & flowers & country looking charming for Tibet.

Said goodbye to others. G.B., MacD & I turned up valley to Raphu – stony at first (tho' carpeted with irises), valley became greener higher up, but wind began to meet us & quite spoilt p.m.

Got into camp at about 12.30, transport 2 hours later.

Feet very bad, so spent a slack p.m. reading. MacDonald caught some snow trout.

Pl. 16 (L)   Noel filming.

8 6/22   *Tsaktsa (13000)*

Marched 8 a.m. Fine & warm but usual d-d wind in our faces all day.

Followed curious narrow valley at a very gentle up gradient for some 8 or 9 miles to pass – lots of birds & some butterflies – no new flowers this side of Doya La (17000).

Sat on pass for an hour or more & sketched [*Pl. 17*]. Descending soon entered delightful country – 3 kinds of azaleas in full flower & in places green turf starred with yellow ranunculus, pink & yellow primulas & several other delightful flowers.

Found a lovely sky blue 'cushion flower'.

Saw many birds but nothing really new. Transport infernally slow & only got into camp at about 6.30 p.m. Pleasant temperature & air at 13000.

But a real jolly day.

Had to wear finneskoe all day as I find both my feet are slightly frost bitten (on ball of foot) thus accounting for awful sore feet I have had for days.

Pl. 17 (L)   From Doya La. 8 6/22.

9 6/22   *Kharta Shikar*

Marched at 8.30 & descended to Arun Valley – at TENG & hence induced transport to proceed to Kharta Shikar – where we arrived at 6.30 & were forced to camp in Jongpen's garden – a d-d nuisance – J. very hospitable.

A lovely day, balmy air; country redolent of flowers & air sweet with songs of birds – many new kinds. I still forced to ride all way as feet no better – collected flowers all day.

Spent an hour or so near TENG & sketched – G.B. & I rode on from TENG to reconnoitre camp.

Kharta valley itself disappointing – stony & not very beautiful. Arun valley however fertile & with lots of trees.

Rose bushes in full flower – marigolds, forget-me-nots, a kind of cowslip, primula, clematis, & many other kinds – the little pink primula starring all green turf.

Butterflies abounded mostly much like ours – saw swallow tail – caught comma, tortoiseshell, clouded yellow, copper & browns of sorts.

10 6/22   *Karta Shikar*

Spent morning riding to reconnoitre for a main camp. Rode 2 or 3 miles up valley – not very beautiful except by side of river which is a glorious snow torrent – then called on Jongpen for an hour. Spent p.m. sorting butterflies & flowers.

Overcast & unsettled looking but delightfully warm & <u>no wind</u>.

11 6/22   *Kharta Shikar*

Pottered up stream after flowers in a.m. Back to dance Tomasha at the Jongpen's at 12.30 – had a bellyful of it. Going again in p.m. After tea tried to sketch but rain again prevented it [*?Pl. 18*].

Getting very bored with Jongpen.

In evening came a letter from Gen. Bruce with sad news of disaster on N. Col. Whole party (4 ropes) swept away by avalanche. 7 coolies killed

– the rest more or less miraculously escaped. The 7 coolies included some of our most gallant – & all were the best of fellows.

Cast a gloom over everything.

Needless to day no further attempt was made – so Everest remains unconquered.

Pl. 19 (L)   Tibetan man and dancing girl, perhaps at Kharta Shikar, 11 6/22.

Pl. 18 (L)   Rain, hopeless. Perhaps near Kharta Shikar, 11 6/22.

12 6/22   *Kharta Shikar*

We all 3 rode some 5 or 6 miles upstream to Hlading (ghastly ponies) – I collecting flowers, MacD. butterflies.

Spent an hour or so there, lunched & sketched – some light rain but most pleasant day.

Back to tea & interview with Jongpen re transport &c. Put the wind up him & hope for better results. Heavyish rain after tea.

13 6/22   *Teng*

Marched about 10 a.m to Teng – all went smoothly – selected pleasant camp near village – refusing 'old Father William's' invitation to camp in his 'Ling'.

Rain came on pretty heavy just as we were settled in. MacD. & G.B. horse coping. I sketched – wrote letters after lunch & strolled after tea. Feet still bad.

14 6/22   *Teng*

Started raining after breakfast & again between 1 & 4 p.m. Rode some 4 miles up Phong Chu valley with G.B. & had lunch there – writing letters all rest of day.

Getting very sick of forced inactivity & want some exercise.

15 6/22   *Teng*

Rain early but cleared & we had a lovely sunny day. G.B. & I rode to gorge of Arun & found a most lovely spot. First a beautiful Cashmere marg & then a real forest of pine & birch swarming with birds & butterflies & smelling sweet of pines, all overlooking a magnificent gorge of the river. Saw any quantity of new or interesting birds. Lunched & pottered & got back to tea.

In morning I bought a pony for Rs130.

16 6/22   *Teng*

Waited in until 11.30 expecting main body but then heard they had not arrived at Tsaksa overnight – so walked two or three miles up Lang Chu – collecting flowers & noting birds. Back to tea.

Pleasant day, feet still useless & can only walk on outside of feet.

17 6/22   *Teng*

Pottered up valley & met main body arriving about 11 a.m.

Back to lunch with them – & spent p.m. in camp with flowers & birds. Lovely day.

18 6/22   *Teng*

Spent a.m. interviewing Dzong Pen. In p.m. watched dancers for a bit & then to junction of Lang Chu & Arun.

After tea busy with mess stores.

19 6/22   *Kharta Shikar*

Camp moved to Kharta Shikar. All of us to my champagne picnic in happy valley – coolies with lunch played us false & I had to retrieve it on pony.

Picnic quite a success.

To K.S. about 5 p.m. – Dzong pen's dinner at 6.30.

20 6/22   *Samchung La*

Marched about 8 – rode & walked up to Chog La, & there met the most delightful scenery & flowers we have yet seen – the little valley of lakes beyond full of new & lovely flowers & the lakes most beautiful.

Sketched [*Pl. 20*] & picked flowers until it came on to rain.

Busy evening sorting flowers.

Pl. 20 (L)   Chog La from below Samchung La. 20 6/22.

21 6/22   *Sakyeteng*

Started early & walked practically whole of a 10 mile march in finneskoe – feet decidedly better.

Samchung La a fine wild pass. Too late for view & only got a glimpse of Chömo Lonzo.

The descent to Sakyeteng through lovely scenery of new type. Two charming lakes & then juniper & pine forest – parts very like Cashmere. Collected many new & beautiful flowers – & saw many new birds.

Camp in a delightful 'marg' amid fine alpine scenery – skinning birds & beasts & arranging flowers in p.m.

22 6/22   *Sakyeteng*

Woke at 4.45 to see Chömo Lonzo – flushed pink in dawn framed in the top of my tent door. Out at 6.30 & sketched [*Pl. 21*]. After bkfast finished skinning stoat – then a scramble through woods with Mallory after flowers.

Came on to rain heavily & spent a hard p.m. skinning & arranging flowers.

Naturalist's job no sinecure.

23 6/22   *Sakyeteng*

Rained all night & all day. Postponed departure for Popti La. Spent whole morning going through flowers & cataloguing them. 223 kinds to date. Ditto with butterflies & moths in p.m. 14 kinds of butterflies, 34 of moths. A dull but useful day.

Pl. 21 (L)   Sakyeteng, Karma Chu. 22 6/22.

24 6/22   *Sakyeteng*

Still drizzling in a.m., raining harder at times, & all day. Mallory & I walked 2 or 3 miles along valley – I wearing boots for 1st time – going a bit tender but great improvement – saw woodcock.

Back to lunch & spent p.m. skinning & arranging flowers.

25 6/22   *Sakyeteng*

Cleared in a.m. & remained half heartedly fine all day with a little light drizzle & lots of clouds & mist.

I walked two or 3 miles up valley through big forest [*Pl. 22*] – looking for birds & flowers. Feet still not much good in boots. Saw nothing very new bar a nutcracker – slack p.m. for a change.

Pl. 22 (L) Juniper tree, Kharta Valley. 25 6/22.

26 6/22 *Sakyeteng*

Coolies' food not arrived so start postponed. Again half hearted weather; lots of cloud & a little rain in p.m.

Spent morning at mess stores & sorting & drying specimens of all sorts. Morris & Noel started for Kyematung & Arun gorge.

In p.m. strolled up nullah & found several new flowers. Later working at flowers &c.

[*On left hand page*]
House Martin
Siskin
White Capped redstart

27 6/22

Marched 8 7.30 a.m. to old camp in Valley of the Lakes. Pretty hard on my feet as too rough to ride. Walked with Mallory collecting flowers all the way. Came on to rain about first lake which was looking most lovely. Reached camp about 3 or 4 p.m. & had a wet evening.

28 6/22

With Mallory towards Samchung La in a.m. – 'my garden' a perfect blaze of primulas, rhododendron & ranunculus.

Found a lot of new flowers. Met one of 3 missing mails.

Busy in p.m. with specimens.

29 6/22

Similar programme. I sketched 'my garden' [*Pl. 23*] – rained rather hard. Mallory down the valley discovered two beautiful new white primulas. Wet p.m.

30 6/22   *Teng*

Wakefield, Mallory, MacDonald & I marched for Teng, 'Ferdy' & Somervell remaining.

Mallory & I decided to go down towards Arun & over pass to E. of Samchung La. Passed his primulas, one a lovely thing growing only in one spot. Went miles down gorge hoping to make 'Happy Valley' but met two bamboo cutters who said there was no way up Arun Gorge so returned some way up hill & by pure luck struck a first class little path over above pass. Made pass about 4 p.m., I very footsore. M. then went on to catch horses waiting for us at bridge below Kharta Shikar.

I followed slowly, met my horse & so home by 7 p.m.

A glorious day over lovely country of varying type – at the lowest point we reached the big purple iris was growing in great beds.

1 7/22   *Teng*

Lovely day with only a sprinkle of rain.

Working all day drying out specimens of all sorts – mending breeches &c. – feet needed a rest.

Strolled a mile or so after tea – 2 missing mails arrived with good news from home.

2 7/22   *Teng*

Morning at mess stores & chasing Kar Singh. In p.m. hair cut & walked up Lang Chu with Mallory after flowers. Found several new bringing total to over 300.

Lovely day again – no rain & perfect temperature.

3 7/22   *Teng*

Mallory, Somervell & 'Ferdie' left for Darjeeling by another route – the two latter to join us at Gantok.

Geoffrey B. & I rode 3 miles with them. Back noon & sacked Kar Singh from mess. In p.m. strolled to Lang Chu.

Lovely a.m., rain after tea.

Morris & Noel fetched up from Arun Gorge which they had followed from Kyematung – a very rough & difficult trip through dense wet forest, continually climbing tremendous hills.

Pl. 23 (L) 'My garden', Samchung La. 29 6/22. Yellow primula. Mauve primula. Marigolds. White rhododendron. Claret rhododendron.

4 7/22   *Teng*

Working at specimens & mess in a.m. In p.m. two Bruces & I rode to mouth of Chung Phu Chu & up hill there. Climbed 2500 feet in an hour, so reckon I'm pretty sound now.

5 7/22   *Lungmé*

Marched about 9. Long march up Arun Valley to junction with Dzaka Chu & then 2 miles up latter to Lungmé. I rode all the way. Horrid wind behind us but a pleasant camp, not an interesting march.

6 7/22   *Dra*

Marched early – another long march – about 18 miles or more. Picked a delightful camp in a little bagh of sea buckthorn – where spent a really pleasant evening. Sketched hard.

An interesting march through fine wild gorges of the river. Colouring improving as we leave the more fertile area & get among the pink & yellow brown paper hills of Tibet proper.

7 7/22   *Ché*

Over the Ché La some 17000 ft. Rather disappointing pass – no view – no flowers – but lovely day & most enjoyable march. Sketched in p.m.

8 7/22   *Shekar Dzong*

Only 8 or 10 mile march. I walked all the way in finneskoe. Loitered an hour or two at bridge over Phung Chu & did two sketches [*Pls 24–5*] – then peacefully on by myself. Had a snack in a charming green upland valley, looked at birds & was absolutely happy.

Found Shekar very different from last time – green turf – no wind – no dust – but people as trying as ever. Had a splendid bathe in water cut.

Pl. 24 (L)   Looking down Phung Chu from under bridge below Ché. 8 7/22.
Pl. 25 (L)   Phung Chu. Bridge below Ché. 8 7/22.

9 7/22   *Kyishong*

15 mile march over rather dull country & a perfectly beastly camp as before here.

I walked all the way in boots – & kept plumb sound – very great relief after nearly 6 weeks.

Weather A.1., warm, very little wind, jolly colours all day.

10 7/22   *Trangsoe Chumbab*

18 mile march – walked about 7 – then rode – dull march with a few charming spots starred with gentians & primulas.

Charming camp on huge green turf meadow by riverside. All bathed – including syces & horses.

Lovely evening – sketched [*Pl. 27*] & lazed.

Pl. 26 (L)   Kyishong. 9 7/22. Too late and hurried.

11 7/22   *Jikyop*

Left Shiling on our R. & came by new route – up Chibling Chu. I walked to Shiling & then rode – dullish march, nice green camp, but beastly wind which spoilt day.

12 7/22   *To*

Walked about 7 or 8 miles & then rode same distance, all up winding river bed over green meadows, sometimes taking to sand hills covered with gorse. Meadows in places simply starred with little pink primulas & tiny white & blue gentians like daisies on an English lawn.

Delightful camp by riverside. All bathed, then I sketched: then heavy shower for ½ hour – lovely colours & reflections. Delightful day – real hot at times & no wind.

Pl. 27 (L)   Looking up Phung Chu from Trangsoe Chumbab. 10 7/22.

Pl. 28 (L)   Tinki Lake. Filthy sketch done in glaring sun. Beautiful effect. 14 7/22.

13 7/22   *Ghangra*

Long march over 17000 ft. pass. Walked some 8 miles & then rode. Fine day, no wind, rain in p.m.

Jolly march with wonderful view over most of Tibet from pass, including lovely great lake N. of Tinki.

Camped in nice wild upland valley. Everyone fished & caught a lot of snow trout. I sketched & worked at flowers &c.

14 7/22   *Tinki*

Short march: I all way on foot – over 15000 ft pass & so into Tinki valley; found this astonishingly lush & green – flowers in places quite wonderful – lots of blue larkspur.

Reflections in lake lovely – sketched [*Pl. 28*] – lunched. Then spent p.m. skinning a rat, packing some birds eggs, labelling mammals &c. After tea along lake trying to identify wild fowl – but, tho' the water was black with them, they wouldn't let me within ½ mile – curious.

Shower in p.m. & dull evening – but no wind, no flies & everything most pleasant – (including most helpful Dzong pen).

15 7/22   *Linga*

10 or 12 mile march – I walked all way bar last mile.

Last April's semi-quicksand, semi-lake has been replaced by a lovely green turf plain – stretching for miles. We were consequently able to shorten march considerably by cutting straight across it.

Fine morning after a wet night. Fine cloud effects culminating in heavy rain in p.m. & a threatening evening – but pleasant & no wind.

I enjoyed march very much – saw some new birds & butterflies.

Arrived in camp, sketched [*Pl. 30*], read English mail of 14th June & so on. Hear family are for Chalet – great news.

We are now on our old tracks again. Extraordinary how country has altered.

Pl. 29 (L)   Undated sketches of an accident en route, perhaps showing the transport officer, Captain Morris.

Pl. 30 (L)   Lingalow looking towards Kampa Dzong. 15 7/22.

16 7/22	*Kampa Dzong*

Rode most of march & found it very dull – Kampa looking charming – sketched – sorted mess stores for Noel – very heavy shower after tea.

17 7/22	*Tatzang*

Marched by route up by nullah behind Dzong. Long march. I footed it to near pass & then rode – going far stronger at these heights (nearly 17000) than when we passed here before.

Pleasant march, but chilly in camp below Tatzang Gampa.

Sketched & sorted flowers.

18 7/22	*Dongka La*

Very long march, & dull. Walked & rode alternately. Walked over 1st pass – 17000 or more – quite easily.

Came on to hail soon after. Camped at a cattle camp – & spent an hour or more in cowherd's tent – before tents arrived.

Very cold & beastly – rained all p.m. This always seems to be a rough march. Spent p.m. skinning voles & sorting flowers. Cheerless & unpleasant day.

19 7/22	*Below Dongka La*

Some 15 or 16 mile march mostly over blizzard country of last April. Dull march, rained persistently towards end. Cold & cheerless camp in open. Busy with butterflies, skinning &c. all p.m.

At this rate shan't be sorry to leave Tibetan Plateau.

20 7/22	*Phari Dzong*

12 mile march into Phari over lovely green plains – a great change from last April. Weather again unsettled – Chomolhari showed up half hidden in clouds for a bit, then rain came on for the afternoon.

Spent a heavy p.m. at specimens & mess stores. Great luxury living in a bungalow again. English mail of 21 6/22.

21 7/22	*Phari Dzong*

Out at 8 a.m. with Parri wallah for ammon – on ponies.

Rode over a fair extent of mountain S.E. of Phari & saw nil except a small herd a long way away. Too many yaks & shepherds.

Back by 3 p.m. & again spent p.m. at mess, &c.

Country traversed full of new (to me) & lovely flowers.

22 7/22	*Gautza*

Macd. & I out at 6 a.m. & into hills S.W. of Phari after ammon. Here again too many cattle grazing – saw nil except some gazelle. So moved on to head of gorge above Gautza for burhel. Stalked a herd of burhel from below (contrary to my ideas) & were spotted of course. Could have fired at a poor head but didn't. Herd crossed onto next mountain.

We followed & again got within about 200x. They got our wind & moved leisurely over crest. I was ready to shoot for some minutes but could only detect good heads with glasses – as soon as I put them down & looked along sights they all merged into one jumble. So never fired. My feet by now d–d sore so Macd. took my rifle & followed over crest &, I regret to say, wounded two without recovering them.

Home by 6 p.m. I too footsore to go on next day so regretfully abandoned last chance of either ammon or burhel – a great disappointment as I had looked forward to this chance.

Pl. 31 (L) Gautza (+ good bye to the plains of Tibet). 23 7/22.

23 7/22 *Yatung*

Sketched for an hour or so from Gautza Bungalow [*Pl. 31*], then walked down this lovely valley among beautiful flowers & the scent of pine woods.

Arrived Yatung late for tiffin – others at Macd's. I followed & spent an hour or so there. Dined with Parsons in 90th Punjabis.

24 & 25 7/22 *Yatung*

Spent these two days pottering up or down valley sketching [*Pl. 32*]. Dined with Macd's on 24th & bought a lot of curios young Macd. had most kindly collected for us. Rain in p.m. 24th.

26 7/22 *Langram*

After an hour saying good bye to our most hospitable friends the Macdonalds, marched late (accompanied by them & an escort of Punjabis to Rinchinggong) & reached Langram in rain at 3.30 p.m. Sketched a bit [*Pl. 33*].

27 7/22 *Gnatong*

Lovely morning – Morris & I early to lake below Kupup. I walked over pass Jelap La without an effort.

Arrived at lake ½ hour too late – thick mist & couldn't even see it. On to Gnatong in pouring rain.

Country marvellously changed since we were here before – a carpet of flowers now.

Gnatong justified its reputation by giving us pouring wet p.m. Comfortable bungalow – good fires.

Good bye to Tibet & very sad to leave it.

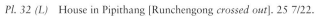

Pl. 32 (L)   House in Pipithang [Runchengong *crossed out*]. 25 7/22.

Pl. 33 (L)   Langram. 26 7/22.

28 7/22   *Sedongchen*

Marched about 7.30 & caught a most magnificent view before the clouds boiled up from below. On one side Kinchenjunga floated high in the air dominating all Sikhim – a cauldron of many coloured clouds. On the other the plains of India spread for hundreds of miles, seen through a rent in the nearer mists.

Rode to Lingtu & then walked down, rain getting heavier as we descended, & air warmer.

The last 500 feet punctuated with leeches of which I picked a score off my clothes. Sedongchen bungalow as pleasant as ever.

Pouring wet p.m.

29 7/22   *Ari*

Marched 7.30 – walked all way to Rangli Chu bungalow – feet suffered badly in process & arrived pretty lame.

Bathed in Rangli R. & then rode up to Ari which was comparatively cool after steamy heat below. Heavy shower as we got in – otherwise fairly fine all day.

Pl. 34 (L)   Kinchenjanga towering over Sikhim from near Gnatong. 28 7/22.

Pl. 35 (L)   Kinchenjunga (mostly hidden) from Lopchu. 1 8/22.

30 7/22   *Pedong*

Walked down to Rangpo Chu again, & again very footsore. Bathed & rode up to Pedong. Dull p.m.

31 7/22   *Kalimpong*

[1 8/22   *Ari] [erased]*

2 8/22   *Lopchu*

3 8/22   *Darjeeling*

THREE LETTERS FROM THE 1922 EXPEDITION

These three letters were written by Norton from Everest to his mother, Edith Norton *née* Wills. She was the daughter of the famous pioneering Alpinist, Sir Alfred Wills. From him she inherited a love of the mountains, and the chalet which he built in the French Alps in the mountains above Sixt. There she presided over family holidays for many years. It was therefore natural for Norton to make comparisons with familiar features of the mountain scenery around the chalet in order to convey something of the character of the Himalayan landscape around Everest.

The letters survive as typed transcripts. Norton must have written other letters home, but these three seem to have been selected for transcription in view of their interest as a record of the first proper attempt on the summit of Everest. Indeed, the second and third letters, together with the diary, provide hitherto unknown information about the pioneering exploration of the East Rongbuk glacier and the 1922 summit attempt with Mallory, Somervell and Morshead. The rows of xxxx at the start and finish of the second letter probably indicate where personal, family details have been omitted from the transcripts: news of his uncle's death and his father's seizure had reached Norton the day before the second letter was written, and must have been referred to at some point. The anticipation of further news from home at the end of the third letter presumably refers to Norton's anxiety about his father. Given his father's illness, and the fact that he himself was still at the time a bachelor aged 38, his mother was the natural person to write to. There are a couple of passing references to his brother Eric, who was always known by the nickname Gen, and his sisters Amy and Gilly.

The transcripts appear to have been typed by someone who had little familiarity with the material. There are some obvious errors in the spellings of some of the names and other less usual words. Some of these have been corrected in pencil on the transcript by a hand which is not Norton's: others have been corrected editorially in the versions which follow.

16500 ft. Camp,
Above Rongbuk.
2.5.22.

Dearest Mummy

We arrived here yesterday and it was only sprung on us last night that a post would leave at midday today – so I must do the best I can, for I have to go off and do a reconnaissance of another camp this morning.

No mail yet. This prolonged delay is inexplicable unless some Dzong Pen is being obstructive – as there must be 3 or 4 mails accumulated somewhere.

We left Shekar on 27th and did a short march to a place called Pangle on the far side of the Phung Chu (Arun). Perfect day – I strolled on ahead and had a most delightful march all by myself with a pleasant loaf of an hour or more by the river. There was hardly any wind this day which always makes all the difference.

Pangle was a tiny village in a kind of Scotch Glen under the Pang La (turf or grassy pass) which we were to cross next day. We were all very happy at getting away from the filth and crowds of Shekar.

Next morning we were off at 7.30. I walked all the way – over the pass, 17000, and down to about the same level the far side. For the first time I went well at this height, reaching the summit of the pass without turning a hair.

Arrived at the pass we were met by the most magnificent view of the whole snow range including Makalu, Everest, Cho Uyo and another magnificent snow peak next it – Gaurisankar, and far to the west Gausamtau.

I did a sketch while, curiously enough, a tremendous argument raged as to whether Everest was or was not Everest. This seems incredible – the reason was that the whole N. face, facing us, was practically a rock peak – almost devoid of snow – utterly different from the east face of which we got so fine a view from Shiling a week ago. There was a certain amount of cloud about which gave rise to the idea that Everest was behind this peak and hidden. However there was soon no doubt.

On this absolutely bare rocky pass (its name 'turf pass' applied only to the slopes below on the N. side) we found a charming gentian like an immature acaulis, a saxifrage, a tiny apollo butterfly, which I caught, and two lizards.

We then descended through a fine limestone gorge (where G. Bruce in front met a herd of burhel quite close – the first game we have seen since Kampa Dzong) to a pleasant camp on a more or less green flat by a river.

That evening it turned stormy looking and very cold – 20° of frost again at night – and next morning a most bitter searching wind which made breakfast in the open pretty trying.

Next morning a longish march of some 18 miles up a river running E. & W. – really the Rongbuk stream – the dullest scenery yet – bare brown rounded hills with snow peaks just appearing over them to S. but mostly obscured by clouds. Got into camp in scenery something like the barest of Scotch hills in winter about 1 p.m. and spent a cold afternoon there.

Next day we marched again in bitter wind about 7.30 and soon swung south on our last lap straight for Everest. We almost at once got into the wildest and most desolate scenery. Practically all traces of vegetation except a little dwarf juniper, sea buckthorn and systus disappeared and the whole valley was a jumble of old moraines and a waste of boulders and rocks, mostly granite, mica schist and limestone whitish in colour. The stream – representing the whole drainage not only of the N. side of Everest but also of Cho Uyo and the intervening peaks and range was not as big as the Giffre below the Chalet.

This is due I suppose to everything being frozen stiff, but also to the dryness of the climate in which all the snow beds and glaciers evaporate straight into the air.

About 1 or 2 p.m. we reached Rongbuk – a very holy monastery in which dwells the most holy lama in Tibet – and camped there – bitterly cold with occasional gleams of sun to warm us for a few minutes.

Towards evening the marvellous peaks and ice cliffs of Everest began to show through the clouds at the head of the valley and for

a few minutes the summit and whole face cleared, giving us a close view of our future campaigning ground – intensely interesting, very beautiful – and, as regards climbing difficulties, rather hopeful.

But the very idea of standing anywhere on those bitter summits now covered with a good powdering of fresh snow was rather impressive.

I sketched feverishly, my water freezing as fast as I put it on the paper, as also my fingers.

Next morning we pushed off on our last trek with yak and donkey transport. The Tibetans driving them were frankly not 'for it', being afraid of devils, and also their animals devoid of forage. After two lightning strikes we finally came to a dead halt some 6 miles up the valley just at the foot of the main Rongbuk glacier from which point they would go no further. We had hoped to get round the corner into the mouth of the E. Rongbuk glacier.

So we made the best of a bad job and established our main base camp in a rocky hollow almost surrounded by old moraine heaps, fairly sheltered and a sun trap in fine weather.

This was last night. Our transport has all gone and from now on we must depend only on cooly transport.

This morning is fine and the sun gloriously hot.

Strutt, Finch and I are just off to decide the position of our first staging camp somewhere at the mouth of the E. Rongbuk glacier.

Then in probably a couple of days Longstaff, Strutt and I carry a reconnaissance to fix two more camps right up to 21000 just under the N. Col. This will mean going light and sleeping one or two nights up the glacier – which will be O.K. if weather holds but beastly in conditions of last 3 days.

I can describe the wildness and grandness of this place with mighty Everest towering over us much as the Buet does over the Chalet.

I must end as I am to be off.

Goodbye my dearest: my dearest love to all.

Ever your most loving son

✼ ✼ ✼

Main Camp,
Rongbuk Glacier Snout,
About 16700
10/5/22.

X   X   X   X   X   X

I have had an intensely interesting time since I last wrote and will lead off by saying I am magnificently fit and well in every respect. I think now almost the only member of the party who has had no diarrhoea, headaches, influenza or other ailment (bar, I think, Morshead).

To hark back. I spent 3rd and 4th busy in this Camp – sorting tents, mess stores &c. for transport to higher camps.

On p.m. of 4th, we had a sufficient fall of snow to make the whole valley white.

On 5th Strutt, Longstaff, Morshead and I started on an expedition to locate 2 more camps up the E. Rongbuk (pronounced Runbu' – the k being silent) Glacier to the foot of the N. Col.

Last year Wheeler went about half way and established a photographic survey point at 19000 odd. Beyond this we trod virgin ground until we joined the tracks of last year's expedition across the head of the glacier from the Lhakpa La to the N. Col. So one has trod ground where probably no man passed before.

We started in glorious weather – the snow already gone by 10 a.m. – plugged up the same route as I described last week and reached No. 1 camp – which we then sited – in time for a mid-day snack. We took 16 coolies carrying light mountain tents, food &c. Our personal kit consisted of eider down sleeping bag (say 15 lbs or 20) on a cooly and a knapsack with change of socks, washing things &c. apiece. I wore my windproof suit, a sweater and No. 2 pair of Carters boots (complete success – I didn't lose one nail). I wore cashmere puttoo hat and didn't feel the sun even on glacier.

That p.m. M. and I reconnoitred route further about half way to No. 2 Camp and so back to our evening meal at 5 p.m.

A party from here had already established a dump of yak dung fuel at this camp and built stone wall 'Sangars' which we roofed with spare tent flies. So we cooked in ordinary way and slept 3 in a sangar.

From now on we cooked with Primus stove – breakfast of tinned bacon, sardines and tea, dinner at 5 or 6 of 'hoosh' compounded of tinned soup, tinned meat ration (Maconochi sort of stuff) and ration biscuits pounded up. Several cups of this make a full meal. Our mid-day meal is chocolate, biscuits, cheese &c.

Next a.m. we all started about 7.30 and plugged continuously up awful rough moraine along the left bank of the glacier – something like the roughest portion of the Guivra stream bed, but much worse going. About 1 p.m we were pounded by a tributary glacier and camped on moraine. By the way from No. 1 camp upwards I never saw a living thing, animal or vegetable until, on our return there 3 days later, one met Alpine choughs and linnets again and the first blades of yellow stunted grass tufts. However on snow at 21,000 I saw tracks of some form of mouse or rat.

In p.m. Morshead and I again reconnoitred and found a way across the snout of this small glacier onto what became the true site of No. 2 Camp about 19000 – a hollow of moraine surrounded on all sides by moraine heaps, hillside or vertical ice walls, 100 feet high. Back to hoosh and sleep at 5.30 p.m. We generally turned in while the sun still illuminated the tops of the surrounding peaks – one just like the Chardonnet, from Lognan – but scarcely ever a 'glow' in Alpine sense either morning or evening.

Longstaff going very badly, only his great courage keeping him going – feeling heart and breathing a lot.

Next morning left L. to build sangars, establish camp, write a report to Bruce and send back half our coolies.

The rest of us went on without coolies and our first job was to cross a much bigger tributary glacier (running down from N. side of N. Peak).

The sun was partially obscured, the weather still and we all suffered from the most pronounced 'glacier lassitude'. This is the worst thing I have met. One feels as if one hadn't a bone in one's body and can hardly drag one foot in front of the other.

If one sits down one can hardly get up again. It is much more pronounced on the glacier surface than on the moraine 10 yards off it. I think it has something to do with lack of evaporation from the skin.

Strutt's glacier knowledge proved useful and we soon found a way across the seracs and so over very smooth slippery ice surface to the far side. Hence we tried to follow the lower eastern slopes of N. peak just clear of the glacier.

We found snow slopes of even 15° to be solid slippery ice harder than anything I have seen (something like the glacier des Dards which Gen. and I crossed on the Floriez) and made slow progress.

Eventually as we approached the corner where one branch of the glacier turns W. to the N. Col we were pounded by cliffs and seracs of a tiny glacier unmarked on the map.

About 2.30 p.m. we gave it up and returned to camp making up our minds to try next day by following the middle of the main E. Rongbuk Glacier itself.

While crossing the junction of the main and tributary glacier Morshead lagged behind and discovered a crossing over the principal obstacle of the main glacier – a great deep ice trough formed in some curious way by the medial moraine.

This proved a very useful move for next day.

Usual procedure in camp. Thermometer registered 2° below zero during the night. I had a grand night in my 7' Farrar flea bag and, of course, all my clothes bar coat.

Next a.m. Longstaff evidently pretty sick. The rest of us with 5 coolies started out again about 8 a.m. It was my idea to utilise these coolies to carry surplus tents and coolies' sleeping bags (8 coolies having already gone down) to No. 3 Camp and so start the establishing of this important camp.

Well, we crossed the mouth of the tributary and the great trough of the main glacier without difficulty. After the cold night the day was grand. Just enough fresh breeze, blazing sun and practically no recurrence of the lassitude of the day before.

I went A.1. all day and felt best as we approached 21000.

But remember one's best at even these heights is a very poor thing. The 'Troupe' will find me an indulgent taskmaster in future!

Once on the middle of the glacier we proceeded up a steady slope of uneven ice lightly covered in places with up to 6 or 8 inches of powdery snow. About 20,000 feet or so, opposite the point we were beaten at yesterday, we encountered good hard snow surface

covering the ice and here we roped as there were small crevasses. Strutt was for forming a camp on the corner of the spur of N. peak but it was decided to try on round the corner and most fortunately we did so, for we discovered an ideal camp (21,000).

This was formed by the moraine and scree at the foot of the bare cliffs forming the S. face of the North Peak.

This spot is so sheltered and such a sun trap that there is a big area of boulder strewn bare ground clear of snow and with lovely water from the melting glacier.

Here we depoted our tents &c. and lay for an hour or so in the sun gazing at our next 'fence', the climb to the N. Col and so on up the N. arête of Everest – rather a thrilling sight seen for the first time at really close quarters.

Strutt and I were agreed that there is quite a hope that there is a route up to N. Col where the ice is snow covered and won't entail excessive step cutting (an awful exertion at this altitude and with such hard ice). But we shall know more of this anon.

Well satisfied with the completion of our job we turned home about 1.15 and retraced our steps without difficulty in about 2 hours. (It had taken us 4 ½ up – one cooly had chucked it and his load been redistributed.)

Arrived in Camp found Longstaff pretty bad. He thought 'flu. Choking and coughing with fever and unable to eat anything solid. Similar symptoms to those some of our coolies have had.

Decided to let him sleep alone in tent to avoid infection. (Strutt and he – M. and I had doubled up previously.) M. and I tossed up who should sleep in sangar, Strutt having been sleeping badly and feeling cold at nights. I lost, roofed sangar with spare fly of tent and spent a restless night owing to being <u>too hot</u> in spite of shedding woollies, socks &c. all night. In the morning the minimum thermometer recorded 5° below zero, so I'm perhaps not such a poor thing at standing cold as Amy and Gilly feared.

Gorgeous sunny morning again. I went so far as to wash though barely able to break ice with huge boulders. It is astonishing how one can do this between sunrise and 10 a.m in this climate, without feeling any chill, though one's hands and face become rather chapped in the process.

I have now a full beard (in fact I have been awarded 1st prize in the beard stakes) and this is a great protection. I pay however in my beak which has been more or less raw for weeks.

Longstaff – largely I believe to save one of us staying up there with him – with his usual courage got up, dressed and expressed his intention of coming down at least as far as No. 1 camp. As he feels altitude a lot it was hard to know whether or no to try and dissuade him. Anyway he slipped off ahead of us and arrived just in front of us as we stayed an hour to depot all surplus food stores &c. and came slowly down building cairns to mark the route.

We thus arrived at No. 1 Camp about noon and found a cook established there, who gave us some lunch.

Hearing others were on their way up to this Camp with coolies and more stores, we left Longstaff who seemed a good deal better. I am very sorry to hear this morning (from Morris, who arrived at No.1 just after we left and spent night there) that Longstaff is worse – fears pleurisy – and a stretcher has been sent for him. He is such a gallant fellow that it is hard to make out how bad he is, but I fear if he consents to come down on a stretcher he is bad – and the road is no peach – something like the worst parts of the lower Salle valley without a road.

We trundled down from No. 1 to here in 1½ hours and so ended our reconnaissance – an interesting and successful trip with each camp established just where we had previously hoped, and route selected and marked with cairns.

I felt last night, and feel this morning as fresh as the day I started. I had a grand night last night in spite of the fact that I left my Farrar flea bag at No.2 pro bono publico and slept in my ordinary bedding which I have brought all this way for this purpose.

19° of frost last night and a cloudless blue sky this morning indicate the continuance of fine weather – but a most unfortunate hitch has occurred. All the Tibetan coolies – some 90 of whom had been promised by the local authorities to carry at least as far as No.2 camp – have offed it back to their villages and I fear we are dependent on our own 40, depleted to a certain extent by sickness.

The next step is for Mallory and Somervell to go up to No.3 and reconnoitre – and, if necessary, cut steps thence to the N. Col where

some sort of a camp is to be established. Meanwhile all available coolies to continuously push stores up from No.1 to No.3.

There is no definite decision yet as to who will then make the first push beyond there – probably without oxygen, while Finch and party prepare to tackle it with.

I find Finch has been in bed 2 days with diarrhoea and looks pulled down. Crawford has it this morning and Mallory – starting for No.3 – is a bit queer inside.

Strutt, just on 50, has stuck this reconnaissance very well and one day (the lassitude day) was perhaps the best of us, but he always sleeps badly and I don't fancy will do much above 23,000.

Morshead is as hard as nails and a very good man indeed tho' not an experienced climber. I keep splendid health but I feel a very poor thing going uphill and am doubtful of my capacity, of course I don't know to what extent others feel the same.

Mallory is the best performer undoubtedly – a sort of Superman at high altitudes as far as I can judge – and Somervell seems to be fittening up to be very good, tho' I didn't think he was going to.

G. Bruce, a great athlete and full of health may go far, I think, but has no mountaineering experience.

Wakefield and Morris are not likely to go very far, I fancy.

This leaves this p.m. so I must cut it short.

A word about the glaciers.

The E. Rongbuk is quite as big as the G. d'Argentière. It collects 8 or 10 tributary glaciers and grinds down through a not dissimilar gorge, apparently an engine of enormous power. Yet at its snout it peters out into a stream about the size of the Guivra stream high up near the Gorge. It seems to disappear into thin air.

Through most of its lower length it is largely formed of huge pinnacles of clear green ice each like a huge sugar loaf. These are not true seracs and their origin is a mystery to me. All snow beds evaporate into the same formation in miniature.

The moraines seem to spring from nowhere in particular. In their upper portion the medial moraines form deep troughs, down almost to the glacier bed apparently – then swelling from no apparent source they form great high roads level with or above the main surface, but far out-topped by the endless procession of sugar loaf pinnacles. As a result the lateral and terminal moraines are vast ridges, terraces and mounds of rock, marking for miles down the valleys the evident and apparently recent shrinkage of the glaciers.

The mountains surrounding Everest appear largely of familiar sedimentary rock, shale, slate and limestone. I should not be surprised if the very top of Everest is sedimentary tho' it is crossed by a band of granite or mica schist above the N. Col. Yet much of the moraine deposit is crystalline – granite and mica schist.

All day (after 10 a.m) and every day the leeward side of all the biggest peaks (so far the S.E.) appear to smoke in white clouds like a banner, giving them a sinister aspect, even while the windward sides and tops are clear.

The daily wind appears to get less as one leaves the plains and up the big glaciers is no more noticeable than in the Alps.

Our coolies are Sherpas (Tibetans from Nepal) Bhutias (true Tibetans, but mostly sophisticated) or Nepalese. In striking contrast to the local Tibetans they are excellent fellows – willing, cheerful and absolutely honest (even the locals are the last, I think). Most of them seem excellent walkers and climbers on rock and none too bad on snow. They offer to go on and climb Everest if or when the Sahibs are 'tired'. I am not at all sure that some won't be competent to do so.

We continue an absolutely congenial party, no friction or trouble of any sort, and so long as we split into various different little parties and then coalesce again I see no reason why we shouldn't continue so, for one thus avoids possibilities of the inevitable friction between small parties long exposed to rough circumstances.

X   X   X   X   X   X
X   X   X   X   X   X

✳   ✳   ✳

Base Camp,
Rongbuk Valley.
25/5/22

My dearest,

This is rather an eventful letter. I will epitomise at once by saying that Mallory, Somervell and I have brought off an assault on Everest (without oxygen) – camped at 25000 – something higher than the highest world's record – and next day climbed to what we make 26,750 just under the junction of N. and N.E. arêtes, thus beating the Duke of Abbruzzi's record by 2000 feet or more. We are now safely back at base camp with no more serious trouble than some not very serious frost bites. We are mostly rather 'done in' temporarily and the intention is very definite that we are not to go up again.

Bruce and Longstaff regard this effort as a great success even if nothing further is achieved. I can at least feel therefore that I have justified my existence.

We had trying weather conditions, and with more luck might have got very near the top.

To hark back:- After the return of our reconnaissance to fix camps, Mallory and Somervell went to No.3 Camp (21000) to arrange a route to N. Col. This was May 10th. On 14th Strutt, Moreshead and I left with instructions to join them and have a go at the mountain with the primary intention of trying to establish a high camp about 25,000.

Well, we marched at 9.30 or so by our previous route – already getting something of a beaten track under coolies' feet – to No. 1 (mid-day) and No. 2 at 4 p.m.

There we slept and next day proceeded to No. 3. The glacier was by now practically sheer ice (nearly all the snow having evaporated) and very slippery. We reached No.3 at mid-day and spent a quiet p.m. I was disappointed to find that the sun leaves this otherwise good camp by 3.20 p.m. As the temperature falls regularly to about 7° below zero there at night, it makes it rather cheerless.

We found M. and S. had reached the N. Col by the route I had anticipated, only encountering ice in 3 or 4 places and had already put up fixed ropes at most of these. They had had a hard day at this and returned to No. 3 quite done.

This ascent to N.Col is a most imposing wall of ice and snow nearly 2000 feet high towering right over one of the heads of the glacier, but except under bad weather conditions (of which more anon) easy enough once steps are cut and ropes fixed. This latter precaution on account of laden coolies of course.

On 16th weather bitterly cold. We all spent a slack morning and worked out plans. In p.m. Mallory, Somervell, Moreshead and self to main Col at head of main E. Rongbuk glacier – looking over into Kama Valley. For some unknown reason I walked as easily as if in Alps – I can't explain why. It was only an hour's walk. The far side of Col is a precipitous drop of rocks and ice on to the vast Kama Glacier. The whole valley was boiling with clouds and mist and looked most imposing with Makalu and other great peaks appearing and disappearing. Perhaps already beginning of monsoon current.

Next day (17th) we all went to N. Col with about a dozen coolies to establish a camp there. We left at 9.15 and arrived on Col at 1.45 after fixing some extra ropes &c. – all quite straightforward snow and ice work – everyone roped. Strutt nearly gave out and only just made the Col, the rest O.K. but there's no blinking the fact that climbing at these altitudes is nothing but pain and grief.

Having pitched 5 little tents of sorts on deep snow but sheltered from the prevalent N.W. wind under a huge wall or cornice of ice, we returned to No.3 in 1 hr. 40 minutes without incident.

You must understand that one lives rather like the beasts which perish at these high camps. One washes very little – in fact above No.3 not at all, as there is no water, one turns in at 5.30 to 6.30 p.m. in all one's clothes and extras, and one's primitive meals are cooked on a Primus or spirit stove; at No.3 a cooly 'cooked' for us, above we did our own (and a miserable job it is).

18th, we again rested with exception of a trial trip with coolies over an ice slope to try some new boots we issued to them.

On 19th we (i.e. Morshead, Mallory, Somervell and self – Strutt having decided he wasn't up to it) started fairly early for N. Col with 10 coolies – lightly laden with a few remaining necessaries for N. Col camp and 4 very light loads to constitute our 25,000ft camp.

We intended to go on from N. Col with two coolies to each load as a reserve.

Reached N. Col without incident and spent p.m. cooking and arranging everything for next day's advance.

By the way, between about 19000 and 25000 we seem to hit off an area of even minimum temperature at night, 5° to 8° below zero.

This cooking is an awful business. Up to 20000 we used Primus stove with petrol or paraffin. Above, a spirit cooker with a patent solid fuel called 'Meta' or absolute alcohol. Both latter are fearfully difficult to ignite at these heights and it takes longer to turn snow into water than to boil the latter once produced. One daren't cook in tent for fear of fire and consequently one sits outside in wind over a stove which certainly has no heat to spare for the wretched cook.

I am glad to say I did at least my share of this wretched job, our food consisted of such easily cooked stuff as Bovril, tea and spaghetti eked out with sweet stuff like chocolate, mint cake &c.

Had quite a good night – snow makes soft lying.

Up at 5 to find 5 out of 10 coolies hopelessly mountain sick (presumably because they closed every aperture of their tents – for their clothes and eider down flea bags were much as ours).

Breakfast consisted of hotting up again water kept over-night in Thermos bottles. Even this an awful slow job.

Finally got off about 8 (?) or perhaps sooner, and proceeded to tread ground where no man has ever been (with only 4 coolies).

It was a cold morning with a wind which everyone seemed to feel more than I did. I can't remember suffering.

Morshead took the lead at first – clad as usual in quite inadequate clothes, e.g. one pair of socks.

I wore Burberry suit, Carters boots with 3 prs. socks and a fine pair of canvas sheep skin lined gloves lent me by Strutt.

We climbed by an easy snow and rock arête without incident until about 12 noon or 1 p.m. By the way it is necessary to halt and camp very early if one is to have any food or anything of a night.

By now wind was horrid and we more or less fled over onto the E. side of the arête and selected the best camp we could find. This was pretty bad – a trifle sheltered but all of such steep rock that we could only find one place even resembling level for a tiny tent. This Morshead and Somervell did their best with, while Mallory and I wasted ½ hour looking for one for ourselves and eventually did what we could with a sloping slab of rock. The coolies helped for about an hour all told – we got two little tents up somehow and then sent them off.

Happily the going had all been easy and they could be trusted to find their own way down. This they did not only to N. Col – where we meant them to sleep, but right away to No. 3 – where we saw them arrive about 6 p.m. They apparently conceived a horror of N. Col.

Afternoon spent in cooking &c. mostly by Somervell and self.

Then came a miserable night. We had two light double eider down sleeping bags sleeping 2 in a bag with no form of mattress. The rocks beneath us were like the pyramids and it was pretty cold and miserable.

In addition I forgot to say that early in the day I had put down my rucksack (stuffed spherical with spare woollies, socks, bedroom slippers &c.) for a moment and it had disappeared with one leap to the Main Rongbuk Glacier below.

Consequently I was none too well off for clothes – tho' of course the others helped me out.

As soon as I lay down I found my right ear was swollen and suppurating – obviously frostbitten, which didn't improve my night as it confined me (and also poor Mallory of course) to sleeping on one side only. I suppose one got some sleep.

Next morning found it snowing hard, and so delayed our start. Somervell and I were both for going on – Mallory, the man of iron, was below par – but readily agreed.

Usual cooking and so on. Started perhaps at 8 – I can't remember.

Morshead after 100x found he wasn't up to it and so returned to camp. I was feeling pretty strong, so led and in fact led 2/3 of day uphill as Mallory wasn't up to it, tho' he more than pulled his weight on the day by leading all way downhill and doing it splendidly.

There is disappointingly little to tell of the climbing. The mountain is simply a great Tenneverge – with very similar angles and type of rock (it is all sedimentary – a great surprise). On this day there were 2 or 3 inches of snow which concealed much and made it rather slippery and beastly even uphill and, downhill, distinctly dangerous at times. On a normal day there isn't, I should say, a bad step from N. Col to Summit provided you pick your way.

I felt no more inconvenience from rarity of atmosphere at nearly 27000 than I have done over several 17000 ft passes and I have very little doubt that had everything gone well – had we camped at say 26000 – got away well – with no snow and struck ideal conditions, we could have walked to the top of Everest without oxygen and without undue inconvenience from rarity of atmosphere.

I am very certain of this in my own mind.

We set 2 p.m. as our upward limit. It sounds rather faint hearted but events justified us.

We had a snack just under (say 400 feet) the N.W. shoulder and gendarme, and started down at 2.30 or so, Mallory leading. He slipped on a nasty snow covered slab once, but I held him and this was his only mistake – otherwise he conducted a most masterly retreat.

We reached our 25,000 ft camp without incident and picked up Morshead. Then our troubles began. M., who had led most of the way up the day before, was almost incapable of retracing his footsteps downhill. He first slipped on a snow slope and, I think, rather lost his nerve temporarily and became fearfully slow apart from his wind – which gave out altogether.

I fancy shortage of drink may have been largely the cause of his trouble.

I slipped badly on an ice or snow slope – steps gave, but no excuse, and carried away everyone except Mallory who belayed very neatly and, in conjunction with Morshead, who landed on an outcrop of rocks, held us up.

After this we crawled down somehow. For several hours I supported Morshead on my right shoulder with Mallory ahead on the rope and Somervell behind, these two fine mountaineers backing up and belaying splendidly all the way. Poor M. whose pluck was indomitable struggled on and on, but the pace was deadly. Finally soon after dark we reached the N. Col.

But the Col is a complicated jumble of cornices, troughs, ridges and some crevasses. From the actual Col to our camp was perhaps 15 minutes by daylight. It was a very different thing to find the way in the dark (even with lanterns) – especially as at this height there was over a foot of new snow.

I can't describe at length our groping on that maze of snow. Happily it was a fairly still night and we got under shelter of a cornice. We ultimately reached our tents at just on 11 p.m.

We then proceeded to dig out some food, and found everything but the spirit cookers (our only hope of drink).

We assumed the coolies had taken these down with them and only found next morning that they were tucked away in a corner of one of the coolies' tents.

One was too thirsty to eat much. I invented a dish which I recommend for similar occasions – strawberry jam and 'Ideal' milk (both solid of course) mixed with powdery snow into a sort of ice. It was really rather good but not very sustaining.

Here we had our proper sleeping bags. Personally I put in a glorious sleep.

Next A.M. we couldn't face breakfast without drink, and as it should not be more than 1¾ hours from Col to No. 3 we decided to breakfast at latter, the discovery of the spirit cookers at the last moment didn't alter our decision, but we had quite failed to reckon with the foot of new snow which had fallen at this point. No slightest sign remained of our original steps and some of the fixed ropes were completely concealed.

The first bit was a nasty steep lip of ice and snow dropping sheer under us with our route running so to speak along the actual lip – a beastly place. I led, followed by M.^d followed by S. and M.^y. I kicked and cut steps for what seemed many hours – each step requiring the most meticulous care – with the feeling that Morshead might go at any moment behind me and trusting to Mallory and Somervell to hold us both if this occurred. The two latter worked liked Trojans behind – belaying and also improving steps for the coolies who would have to come up later to get our bedding &c.

To make a long story short we reached the glacier below without incident, in, I suppose, 3 hours.

Here we met Finch, G. Bruce, Wakefield and a party of some 20 coolies carrying oxygen cylinders to start a series of depots.

They gave us <u>tea</u> and <u>brandy</u> out of Thermos bottles. I wish I could describe what it tasted like.

We then staggered into No.3 Camp, and were treated in the most delightful way – poor old Strutt (who felt badly not having been able to come, I fancy) going down on his knees to take off my boots and put his moccasins on my feet while I fed – Wakefield (who turned back with us) giving me his flea bag all the afternoon and so on.

Wakefield proceeded to count the damage.

Apparently Morshead had every finger of both hands and one toe (tho' this was only discovered next day) pretty badly frostbitten. Mallory had most of his fingers frostbitten (and one toe touched, I think). Somervell has only one finger a bit touched and my right ear is frostbitten – nothing to worry about – much like Gen's was – ski-ing – I expect.

The oxygen party got back that p.m. Finch and G. Bruce had used it all the way to Col. and report it a triumphal success, making them go like birds at the time and feel A.1. for 2 hours after.

Well, I was very tired that p.m. – I slept like a log and had a magnificent breakfast next a.m. We decided to try and come right through to Base Camp on account of these frostbites – and we did it, but only once before (in 1914) have I known what it is like to be so exhausted. Glacier lassitude was in full blast between No. 3 and No. 2. From No. 2 to No. 1 I bucked up a lot, but between No. 1 and Base Camp the trough beside the glacier which one follows for 2 or 3 miles had been invaded by a stream – owing to bursting of a lake in the glacier – and one had to work out a new route over boulders and scree – the Salle Valley touch again – and this was about the limit after a hard day. However we all came into camp stepping up to our chins at 5 p.m. and got a great welcome – whiskey in our tea, fizz and quails in aspic for dinner, and then the luxury of a full sized 80 lb. tent and bed.

Longstaff says we were an awful looking sight – all covered with bandages with all the skin of our faces frostbitten and so on.

But now that one has had a day to clean up there's not much damage. The only doubtful case is Morshead and we are all pretty anxious about his fingers and fear he may lose some.

His toe he only discovered on the way down here from No. 3, but I don't think it is going to come off.

I spent all y'day doing absolutely nothing and all this morning writing this. Yet I'm still an absolute rag, as are Mallory and Morshead and Strutt (tho' 2 latter haven't done as much).

Somervell is probably the best off; he is very strong and I fancy has great recuperative power.

Later – after lunch.

A note has come down from No.3 (where are Finch, Wakefield, Noel, G. Bruce and Crawford) from Noel to say that he has transported all his photographic outfit to the N. Col and is ready to take Finch's and G. Bruce's 'great attempt'.

I gather this means a definite attempt but with a limited number of bottles – not enough, to my mind, to reach the top.

Whether or no this means that they are going to reserve enough oxygen for another full dress attempt I don't know.

If so, I don't know who will be available for it – Somervell now talks of trying to go up again almost at once – dependent on Longstaff's examination of his heart. I confess I am not yet up to it – if only on account of my ear. Mallory certainly isn't.

Finch doesn't think Crawford or Wakefield is up to it – or suitable.

In any case the matter will shortly be decided one way or the other. It is necessary to give 14 days notice for transport to leave here – and this has been given to-day.

When we leave here (June 10th or 11th), the following will probably be the procedure.

Strutt, Longstaff, Finch, Mallory, perhaps Morshead – perhaps I – will return at once the way we came or perhaps a rather shorter route through Sikhim.

The remainder will play about for say 3 weeks high up in the Kharta and Kama Valleys – photographing, collecting flowers &c., and recuperating from too great altitudes to a pleasant land of moderate altitudes – forests, wood fires and vegetation.

This party should be back in Darjeeling by August 15th or 20th and with luck one might be home by about 1st week in September (i.e. 2nd or 9th).

My intention is to remain with the second party, with the reservation that I shall join the first or even come through on my own just before or after them if I hear any news from home before then to induce me to do. Another possible factor might be my teeth. In common with several others I have been losing bits of stopping out of them, and tho' none have given me any trouble so far, they may begin to do so – in which case I shan't risk staying too long away from civilization.

Well, so much for the moment, I have another letter of yours and one of Critleys to comment on and will refer to them before I send this off. I may have another home mail by then but I wanted to break the back of all this long account first.

Please excuse writing, my fingers are full of cuts and chips into which the frost has got.

<u>*26/5/22.*</u> *I am picking up fast and shall be myself in a day or two – in spite of the infernal cold here – which is either worse than ever, or else I feel it more. The wind whistles through this camp all day and tho' there are a good many hours of sunshine, it is impossible to get in sun and out of wind; I am afraid monsoon conditions are beginning. The weather over the mountain looks daily now like a conflict between the monsoon current up the Kama Valley and the usual N.W. wind in the opposite direction – and I fear there is a good deal of fresh snow on the mountain. There is no comfort here and we shall all be glad to go, tho' whether there is such a thing as warm weather anywhere in Tibet at any season appears doubtful.*

My ear is rather a nuisance and kept me awake much of last night but this is, I suppose, quite normal.

EVEREST 1924

On the day he returned to Darjeeling at the end of the 1922 expedition, Norton was already contemplating another attempt on Everest the following year. With fair weather, he wrote to a friend, 'I can't see what is to stop a stout party *without oxygen*, and oxygen will be improved and also do the trick, I think.' As he sailed back to Europe, he wrote to the Mount Everest Committee offering his services. He disembarked at Marseilles and travelled straight to the family chalet above Sixt. When he reported back for duty at the War Office, he was immediately posted to Chanak in the Dardanelles, where British forces were in a tense confrontation with the Turks. He therefore missed the public meetings and lectures and the screenings of Noel's film which occupied some members of the 1922 expedition over the winter and into 1923. He was also unable to write his intended botanical contribution to the expedition book, which appeared during the course of 1923. It was perhaps fortunate for him that it proved impossible to organise another attempt on Everest that year, since he remained at Chanak till August 1923.

At the end of the 1922 expedition, General Bruce had reported that Norton was 'the great success of the expedition ... a first-rate all-round mountaineer, and full of every sort of interest'. He was

therefore an obvious choice for the 1924 expedition, which he was invited to join in November 1923 as second-in-command to General Bruce and as climbing leader. While delighted to join the expedition, he accepted the deputy leadership only with reluctance, protesting that in his opinion Mallory was much better qualified for the appointment. Nonetheless, he set about making preparations immediately. Sailing out to Bombay with the General in February 1924, he missed the final act of the 1922 expedition. During the first Winter Olympic Games, which were held at Chamonix in January and February 1924, the members of the 1922 expedition were awarded special gold medals for *alpinisme* for their endeavours on Everest. The medals were accepted on their behalf by Colonel Strutt during the closing ceremony of the Games on 5 February. At the expedition's request, two further medals were subsequently awarded to the porters who had played such a crucial role in the attempt on the mountain.

Norton reached Darjeeling with General Bruce at the beginning of March. In between attending to the expedition stores, he attempted to capture in paint the view of the snow-capped Himalayas seemingly floating high in the air (Pls 36–8). Gradually, the members of the party assembled. The other veterans from 1922 were Mallory, Somervell, Geoff Bruce and Noel, the last of whom had been instrumental in financing the new expedition. They were

Fig. X   The 1924 expedition members at Darjeeling, March 1924, with Norton standing in the centre.

joined by Noel Odell, a geologist from an oil company operation in southern Persia. A member of the Alpine Club, he had been invited to join an Oxford University expedition to Spitsbergen in 1923, where he had met and been impressed by Andrew Irvine, a young sportsman with a bent for tinkering with machinery. Irvine was selected on Odell's recommendation and was to put his practical skills to good use on the oxygen equipment. John Hazard and Bentley Beetham were both experienced mountaineers, while Edward Shebbeare brought many years' experience of India and knowledge of local languages to his duties as transport officer (along with Geoff Bruce). Richard Hingston, an RAF medical

officer based in Baghdad, doubled as expedition doctor and naturalist. He had a particular passion for insects, something that was captured in one of Norton's most amusing sketches (Pl. 51). John Macdonald, who met the expedition at Chumbi, was not officially a member of the team, but was active in their support all through Tibet. Of the local people who are mentioned by name in the diary, a position of special honour is held by the four 'Tigers' or high-altitude porters who helped establish the highest camp on Everest: Nuboo Yishay, Llakpa Chedi, Semchumbi and Lobsang Tashi. Without their efforts the summit attempts would not have been possible.

The approach march

The approach march stuck closely to the route followed on the return from Everest in 1922. As deputy leader, Norton took charge of the second party which left Kalimpong on 28 March. The early diary entries again reflect Norton's fascination with the flora and the birds sighted en route, and the easy and congenial relationships among the expedition members (including a practical joke which fooled the second party on April Fools' Day). The two parties met up temporarily at Yatung, where the entry for 3 April records the first concerns about General Bruce's health. Norton took over the leadership of the first party out of Yatung so as to allow the General to follow on with the second party after an additional day's rest. When the two groups met up again at Phari, the General was no better. It was decided that he would divert to a more easterly route out of Phari, accompanied by Hingston and John Macdonald. By travelling north in the footsteps of the 1921 reconnaissance expedition to Tuna and then westwards to Tatsang, he could avoid the severe conditions over the Donka La. So from leaving Phari on 7 April, Norton was effectively in charge of the expedition (which from then on travelled out together). After several anxious days, the news reached him at Kampa Dzong on 13 April that General Bruce had been pronounced unfit to travel further and was turning back, accompanied by Hingston. Leadership of the expedition thus passed definitively to Norton, who noted, in one of the shortest entries in his diary: 'So I now carry on' – or, as we would say nowadays, 'take over'.

From Phari onwards, mentions of wildlife and flora largely disappear from the diary, to be replaced by the concerns of leadership. Periodic negotiations were required with the Dzong Pens and Gembus of each administrative centre they passed through, sometimes resulting in wrangles and delays over transport and porters. Norton took over responsibility for a series of dispatches for *The Times*, one of the sponsors of the

Fig. XI   Norton (left) and Geoff Bruce en route near Gautza in the Chumbi valley, probably 3 or 4 April 1924.

Fig. XII   General Bruce (centre) and Norton (in pale raincoat) engaged in discussions about transport at Phari, 4–7 April 1924.

Fig. XIII   Breakfast at Lung dôk, 10 April 1924, Norton centre-left carrying a sack.

expedition, which were published in London two or three weeks after they were written. After composing the second, he confided to the diary: 'fearful tripe – I only hope the British public will like it' (20 April). There were anxieties over the health and fitness of the party. Beetham contracted dysentery, from which he never recovered sufficiently to take an active part in the climbing. Mallory also gave cause for concern. Norton himself was going well, for the most part, though on 27 April, as they approached Everest, he noted some concern about his eyes.

As they travelled westwards across the Tibetan plateau, there were regular practices with the oxygen equipment, and frequent discussions with Mallory on the strategy to be followed on Everest. When he took over as expedition leader, Norton appointed Mallory as his deputy and as leader of the climbing party, and he insisted that Mallory, not he himself, should take any decisions as to his own inclusion in any summit attempts. The two men had been discussing the climbing strategy for months, inconclusively. Eventually on 17 April, Mallory came up with a proposal which met with Norton's approval and was put to the rest of the team. Final logistical plans could then be put in place,

but the oxygen equipment was a major worry: 'the apparatus is failing all along the line – 33 cylinders out of 90 have leaked – almost every instrument leaks at every joint & tube. But for the mechanical genius of Odell & Irvine we should be hopelessly done' (21 April).

Hanging over the entire expedition was the spectre of the weather. Anxiously surveying the stormy, windy conditions for signs of an early monsoon, Norton wondered 'what *is* the weather doing?' (21 April). His forebodings were soon to be realised.

The struggle to reach the North Col

The plan of campaign envisaged re-establishing Camps I, II and III in the same locations along the East Rongbuk glacier as in 1922, and Camp IV at the top of the mighty ice cliff, by the North Col. Three high camps (V–VII) were then to be placed at 25,500ft, 26,500ft and 27,300ft, and on the appointed day two parties, of two climbers each, would set out for the summit at the same time. One party, without oxygen, would start from

Fig. XIV   A relaxed group photo during the meeting with the Dzong pen of Shekar Dzong, 24 April 1924. From left to right: two Tibetan officials, Mallory, Norton, the Dzong pen, Geoff Bruce.

Fig. XV   Expedition members photographed at Base Camp in late April or early May 1924. From left to right, seated: Shebbeare, Geoff Bruce, Somervell, Beetham; standing: Irvine, Mallory, Norton, Odell, Macdonald.

Camp VII, while the other, using oxygen, would set off from Camp VI. The two parties would hope to meet at the summit, but if one party got into difficulty, the other would hopefully be able to come to their aid.

Base Camp was pitched on 29 April in bitterly cold weather. Supervising operations for the first week from Base Camp, Norton was initially able to contemplate progress up the East Rongbuk glacier 'up to time beyond my expectations' (2 May), in spite of a strike by many of the local porters and the withholding of supplies by one of the local officials. But from the start of May the diary ominously notes poor weather high on the mountain. On 7 May, Norton went up to Camp II and 'found things far

from well', with forty porters sleeping in a camp designed for twenty, forcing him to break into the stores which had been specially prepared for the high camps. Uncertain as to the cause of the confusion, he noted that 'our shortage of transport officers & of people who can really talk the language [of the porters] is going to be a very severe handicap'. From then on the situation deteriorated rapidly. Many loads had been dumped below Camp III, leaving it inadequately provisioned. On 8 May, Odell and Hazard failed in an attempt to reach the North Col. On 9 May, Norton, going up to Camp III to assess the situation, found himself caught in a terrible blizzard, followed the next day by appalling wind on the glacier, and a restless night in a tent 'shaken like a rat by a terrier' (10 May). With the entire expedition in disarray and the route to the North Col unsafe for days to come, on 11 May Norton ordered a complete evacuation back to Base Camp. As he retreated down the glacier, he found several of the porters and Gurkhas dangerously ill. Lance Naik Shamshar had to be carried down from Camp I insensible, and died before reaching Base Camp. Thus the first attempt on the mountain ended without even reaching the North Col.

Back at Base Camp, Norton took stock of the situation, confiding to his diary that the retreat was 'very bitter'. A revised plan of campaign, based on the original plan, was drawn up; and on 15 May almost the entire expedition descended to the Rongbuk Monastery to receive the blessing of the head Lama, a highly revered figure who also impressed the climbers. As a measure to restore the morale of the porters, it succeeded entirely.

The next morning brought clear weather and an immediate start up the East Rongbuk glacier. This time Norton himself, with help from Mallory, pioneered a new route up the great ice wall above Camp III to the North Col. With the disaster of two years earlier very much in his mind, it was designed to be safe from avalanches. On the way down again the same afternoon, Mallory descended by the 1922 route and got into serious trouble on his own. From now on, Norton's anxieties about the weather and the dangers of the whole enterprise feature in the diary every day.

Fig. XVII   *During the ceremony of blessing at the Rongbuk Monastery on 15 May 1924. From left to right, seated: Geoff Bruce, Mallory, Norton; behind, porters queuing to be blessed by the head Lama.*

Fig. XVI   *A cheerful group portrait apparently posed in front of a sangar at Camp I or II, probably on 6 May 1924, shortly before the set-back on the East Rongbuk glacier. The men seated in front are perhaps the Gurkha NCOs who established Camps I and II; standing behind, from left to right, are Geoff Bruce, Mallory, Norton, Somervell.*

The night of 22 May saw the lowest temperature ever recorded up to that date on Everest at –24°F (–31°C). The following day it emerged that four of the porters were marooned on the North Col, while heavy monsoon snow had made the route above Camp III 'palpably unsafe'. With the porters generally in a poor state, and all his fellow climbers more or less unwell, Norton was forced to order a second retreat, while preparing for a mission to rescue the stranded porters the next day. The rescue party consisted of himself, Mallory and Somervell. Geoff Bruce was left out so that he could take command of the remnant of the expedition should the mission end in disaster, as seemed all too likely. In the event, against all the odds, the porters were rescued, largely thanks to magnificent mountain-craft by the ice-cool Somervell; but Norton's frustration as he descended the East Rongbuk glacier once more in terrible weather is clear: 'The weather is cruel: with 2 years ago's weather I believe we should now be homeward bound with the mountain defeated.'

The summit attempts

There was now only a remote chance of reaching the summit. Not one of the high camps had been established, Camp IV at the North Col had not been properly supplied – not a single oxygen bottle had reached it yet – and stores were scattered the length of the East Rongbuk glacier. Few of the porters were capable of high-altitude work, and the climbers were already for the most part exhausted. However, on 27 May there was a break in the weather. The previous climbing plan was abandoned. Instead, it was decided to mount two successive attempts on the summit without oxygen, launched from two high camps above the North Col, rather than three. The first was to be by Mallory and Bruce, the second by Norton and Somervell.

They set off one last time from Camp I on 29 May. On 1 June, with the weather still fine, but very windy at altitude, Mallory and Bruce with eight porters left Camp IV and established Camp V at over 25,000ft. They spent the night there with three of the porters, with the intention of carrying on up to establish Camp VI in the morning. The next day Norton and Somervell climbed up from Camp IV in their turn, and were surprised to meet the first party coming down again. They had encountered exceptionally bitter and debilitating winds the previous day, and not even Geoff Bruce, who had a special rapport with Himalayan people, had been able to persuade the porters to go any higher. Norton and Somervell spent the night at Camp V with four porters. On the morning of 3 June, Norton finally managed, with enormous difficulty, to persuade three of the porters to carry on as far as Camp VI, which was pitched at a height of about 26,800ft. From there they set out on their attempt on 4 June, in brilliantly fine and calm weather, but bitterly cold. At about 28,000ft Somervell, who had been struggling with a severe high-altitude cough, was unable to continue, but Norton carried on climbing solo. He eventually reached an altitude of 28,126ft (8,572m) before turning back.

Fig. XVIII   Norton seated high on the north face of Everest photographed by Somervell during their summit attempt on 3–4 June 1924.

This remained for 54 years an unbroken world altitude record for mountaineers not using oxygen. The two men reached Camp IV after nightfall, and soon after Norton was struck with severe snow-blindness. He was blind and in great pain for three days.

The next day, as he lay in his tent at Camp IV, sightless and in agony, Norton was approached by Mallory with a plan for a final summit attempt with Irvine, using oxygen. He crawled out of his tent to urge the porters to go up one more time. On 6 June he bade farewell to Mallory and Irvine as they started for Camp V (he was literally unable to see them off), and he decided, blind though he was, to go down to Camp III and await developments there. The 8th was the day of Mallory and Irvine's summit attempt. The diary records the increasing anxiety as the men at Camp III trained their eyes on the mountain. By late morning on the 9th Norton concluded that disaster had probably struck, and gave orders accordingly. 'Of all the truly miserable days I have spent at [Camp] III this [is] by far the worst.' In the absence of any radio communication, it was another whole day before confirmation was received that Mallory and Irvine had disappeared.

It only remained to shepherd the remaining men down to Base Camp and abandon Everest. Though still far from recovered from his own ordeal, it fell to Norton to inform the expedition sponsors in London, and the world at large through his despatch to *The Times*, of the loss of Mallory and Irvine. He also wrote letters of condolence to Mallory's widow Ruth and Irvine's father, which are printed below. One final task was to erect a large cairn in memory of all those who had died on the three Everest expeditions. Norton found time to sketch it before leaving Base Camp for the last time (Pls 78–9).

Fig. XIX   Norton photographed by Somervell on 4 June 1924, climbing solo above 28,000ft on his way to establish a record height without supplementary oxygen, with the summit of Everest in the background.

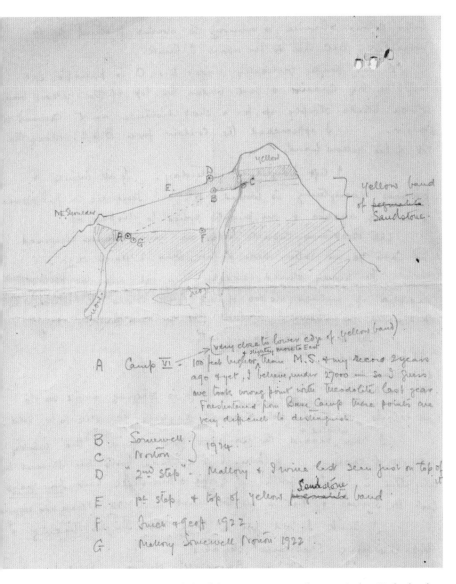

Fig. XX   Annotated pencil sketch by Norton sent in a letter to Arthur Hinks dated 14 August 1924, showing the points reached by the different summit parties in 1922 and 1924. The sketch is based on Norton's watercolour from Chogorong (Pl. 81) and from memory. He believed that it gave a much fairer impression of the upper reaches of the mountain than the photograph used as the basis for several widely reproduced diagrams showing the heights reached by the various parties. The photograph had been taken from a much lower level, and the resulting foreshortening, he believed, gave a very erroneous idea both of the steepness of the mountain and also of the relative heights of the points reached by the different climbers.

The rest period and the return

As in 1922, the expedition retreated to lower altitudes to recuperate before starting on the return march, apart from Hazard and Noel who departed separately. This time they headed west towards the Rongshar valley, another of the great rivers that carve their way down from the Tibetan plateau through the Himalayan range into Nepal. On 16 June they headed north-west from Rongbuk. The first night, spent at Chogorong, yielded a 'wonderful view of Everest' which Norton painted to good effect (Pl. 81). They crossed a pass and descended to Kyabrak, a 'tiny squalid village' lying on the ancient trade route between Sola Kombu, in Nepal, and the Tibetan highlands. From here their route led south-west over another pass, the Pusi La, beyond which start the head-waters of the Rongshar.

They descended the Rongshar valley, which rapidly became a deep gorge, and set up a base at Tropdé. The weather was now full monsoon, with mist, cloud and rain shutting off views of the high peaks, in particular Gaurisankar, a famously beautiful mountain they were all keen to see. The warmer, balmy temperatures helped their recovery, though Norton was suffering in his feet much as he had in 1922. He was generally disappointed with the spring flowers compared to those in the Kama valley two years earlier, but produced some delightful flower sketches all the same (Pls 84–5). Five of the party pushed on down the Rongshar gorge on 24 June, on a four-day excursion that crossed the border into Nepal before returning. In their absence, Norton and Somervell climbed above Tropdé and found a high campsite from which they achieved splendid views of the snow peak of Gaurisankar, much to the envy of the returning group.

On 30 June, they started the journey home. From this point onwards right back to Darjeeling, the diary entries become very brief. After crossing the Pusi La back to Kyabrak, they headed north to Tingri and then east to Shekar Dzong. This was new country for Norton, though the reconnaissance expedition had come this way in 1921. From Shekar Dzong, the route was for the most part identical to the outward journey. With the monsoon in

Fig. XXI   *Norton (left) and Somervell photographed at Base Camp after their summit attempt, with the three Tigers who helped establish Camp VI, Nuboo Yishay, Llakpa Chedi and Semchumbi, 11–15 June 1924. Norton is wearing finneskoe on his frostbitten feet.*

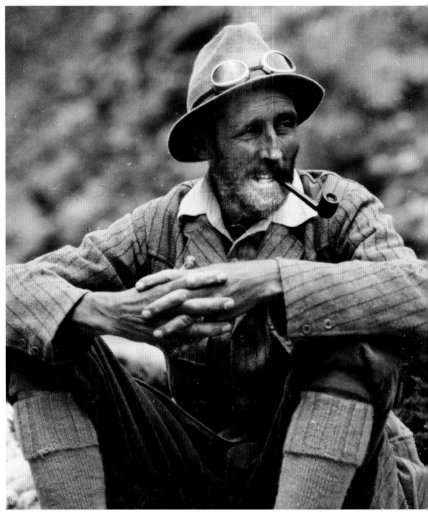

ABOVE: Fig. XXIII   *An informal portrait of Norton, probably taken at some point on the return journey.*

LEFT: Fig. XXII   *Norton on horseback in front of two large structures faced with mani stones. Presumably on the return journey, since he is wearing finneskoe on his feet.*

OPPOSITE: Fig. XXIV   *The expedition members on their return to Darjeeling, standing from left to right: Hingston, Hazard, Norton (in pale raincoat), Beetham, Geoff Bruce, Somervell, General Bruce.*

full swing, they travelled through persistent rain. After Yatung, they diverged from the outward route, crossing the Tibet/Sikkim border by the Nathu La rather than the Jelap La. At Gangtok Norton paid a courtesy call on the Maharajah of Sikkim. For most of the final week he was forced to ride rather than walk because of blood poisoning in his left foot. But on 1 August they were met with motor transport by General Bruce, who had recovered from his illness, and travelled the last 6 miles back to Darjeeling in style.

THE 1924 DIARY AND SKETCHES

26 3/24   *Kalimpong*

Left Darjeeling 6.45 a.m. in motor with Mallory & Hingston & drove to 6th Mile Stone: thence walked to Peshok Tea Garden & had breakfast with Mr & Mrs Lister about 9 a.m., Somervell joining us ½ hour later. Delightful spot swarming with bird & insect life – Hingston fascinated & wants to spend 2 years there. Left again 9.50 & walked by short cuts to Tista Bridge where we met ponies & rode up to Kalimpong D.B. arriving about 1.30. Spent whole p.m. with stores &c.

27 3/24   *Kalimpong*

Same tamasha at Dr Graham's homes as last time. As before much impressed by the whole show. After this 1st party (2 Bruces, Somervell, Beetham, Hazard, Noel & Lloyd) left for Pedong with 40 mules. We returned to D.B. & worked until tea time at stores &c. Tea with Mrs Macdonald & the Perrys, dinner with the Waights (subdivisional offr).

28 3/24   *Pedong*

Breakfast with Dr & Miss Graham – a very pleasant send off. Meanwhile Shebbeare got all the mules (51) off & we left about 10 a.m. Rode nearly to pass & then walked to Pedong arriving 1.15.

In p.m. Shebbeare & I to call on Ranger. After worked at store book until dark.

2nd party consists of Mallory, Hingston, Shebbeare, Odell, Irvine & self.

All vestige of a view completely obliterated by haze & smoke from burning forests in Tista Valley.

Pl. 36 (L)   Kinchenjunga from Darjeeling. 9 3/24. Mountain more golden. Valley more hazy.

Pl. 37 (L)   Kinchenjunga from Darjeeling. 10 3/24. Mountain should float in mid air.

Pl. 39 (S)
Hazard. Noel.
Geoff Bruce.
Mallory.

ABOVE: *Pl. 38 (L)*
Kinchenjunga from
Darjeeling. March 1924.

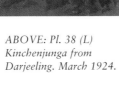

LEFT: *Pl. 41 (S) Two*
profile sketches, perhaps
of one of the European
settlers in Sikkim, and
a man in a hat, thought
to be Karma Paul, the
interpreter.

RIGHT: *Pl. 40 (S)*
Donkey and profile head.

29 3/24   *Rangli*

Marched about 8 – walked to bridge – stream very much smaller than last year but equally beautiful.

Then rode to Rhenok; thence by lower road to near junction of Resi & Rangli Chu; thence walked the 6 or 7 miles up the L. bank of the stream to Rangli, arriving 1.30. Sun not very strong & overcast in p.m. so had a cool & pleasant march through very charming country tho' hardly as pretty as I expected the lower road to be.

After tiffin sat in stream bed with Hingston, looking at birds – Cinclus Asiaticus, the white-capped redstart, forktail Himalayan pied kingfisher, common kingfisher & a puzzling wagtail. Sketched [*Pl. 42*], returned to bungalow & worked at stores for an hour or so. Heavy hailstorm at dusk.

All going very smoothly so far.

[*On opposite page*]
Wagtail –
olive green crown, nape & back; breast ditto; rest of under parts yellowish-white becoming lemon yellow towards flanks & vent, wings darker than back, apparently barred lighter supercilium – legs pinky flesh colour. Tail apparently olive with marginal white feathers.

Pl. 42 (L)   Rangli Chu. 29 3/24.

30 3/24   *Phadamchen (or Sedongchen)*

Marched about 8 – walked for a mile or so & then rode to ½ way village; thence Mallory & I walked, arriving about 11.20. Pleasant cool clear day, hottish sun; views from Bungalow (which we have never seen before on account of haze or mist) disappointing. No snow visible. After an A.1. lunch strolled along contour path through forest, sat & wrote home & so back to tea. Temperature just right in sun.

After tea finished store book – did some Nepali.

We are a pleasant congenial party & so far it has been a most glorious trip.

Pl. 43 (L)   R.W. G.H [*ingston*] in Morris's suit. *Bearded head, probably Shebbeare.*

Pl. 44 (S)   Kinchenjunga from near Lungtu. 31 3/24.

31 3/24   *Gnatong*

Marched at 7.30 – I climbed about 3000 feet & then sat & sketched gorgeous view of Kinchenjunga with brilliant red rhododendrons in the foreground [*Pl. 44*]. Then rode up to the little tea shop at Langtu. Then walked by N. side of hill to the col & thence walked & rode alternately, arriving 1.15.

As before found this the most delightful march of the lot. Magnolias & the red rhododendron in full bloom – higher the mauve primula in beds of colour. On the plateau there was a lot of freshish snow, & it was pretty chilly tho' warmer than last year.

After lunch issued coolies' blankets &c – painted a bit [*Pls 45–6*] & then did high altitude stores with Mallory until dinner.

On way up saw a flock of siskins, lammergeyer, a bronze green laughing thrush I don't know, Indian redstart & some pipits & finches I couldn't identify.

A happy day.

Pl. 45 (S)   Lungtu Mount[n]. 31 3/24.

Pl. 46 (S)   The Gnatong primula. 31 3/24. Primula denticulata.

Pl. 47 (L)   Chomolhari from Jelep La. 1 4/24.

1 4/24   *Langram*   **2 4/24**

Having decided to march over the Jelap & camp at Langram instead of halting at Kupup, we marched at 7.50.

I rode to Kupup (9.30) & thence Mallory & I walked. Made the Jelap in 1¼ hours (10.45) without turning a hair. Very little wind & pleasant enough to sit for ¼ hour or more on the pass. View of Chomolhari & Cho Traké absolutely perfect. Much fresh snow obviously on the Tibetan plateau (2½ feet a day or two earlier we heard subsequently). Descended 500 ft. & sketched for an hour [*Pl. 48*], & then on to Langram bungalow by 1.30.

Pleasant loafing p.m. in jolly jungly bungalow.

Geoff left us a ½ bottle of whisky – which turned out to be tea: we failed to connect this with April 1st.

[*On opposite page*]
Saw lammergeyer, Hodson's rose finch, Himalayan accentor?

Marched 7.30: I on flat feet all the way by myself – reaching Yatung 11 a.m. A really delightful march which I enjoyed immensely: birds of all sorts – smoky willow warbler, Nepalensis wren, plumbeous & white capped redstart, Cornish chough, tree sparrow. Saw the Gnatong primula besides any number of denticulata & the little daphne. In p.m. J. Macdonald gave us a tamasha – Tibetan dancing & singing & lots of chang & arrak – too much for most.

Pretty chilly in shorts in p.m.

Pl. 49 (S)   Dancing girl, Yatung. Also pencil sketch of a chough. Probably 2 4/24.

Pl. 48 (S)   Langrang. 1 4/24.

Pl. 50 (S)   Girl in local costume. Profile head probably of a member of the expedition. Rear view of Andrew Irvine (identified by Julie Summers). Probably at Yatung, 2–3 4/24.

3 4/24 *Gautza*

It having been decided that the General should spend one more day at Yatung, I changed to 1st party.

Got mules off by about 9; we stayed to lunch before marching. Went for a stroll with Hingston looking at birds & insects until lunch.

After lunch started with Geoff at 2 p.m. on my new pony (bought from John Macdonald) – rode the greater part of the march, walking only 2 or 3 miles.

Thought the valley more delightful than ever – the golden stems of the birches imparted a glow to the forest like sunlight.

People all most friendly – putting out their tongues in the most engaging way.

Left behind all flowers except the daphne about Galinka.

Arrived Gautza at 6 & had a slap-up tea.

4 4/24 *Phari*

Marched about 8 & walked to the frozen waterfall. There we had tea in the 'Pari wallah's' basti: I picked up a herd of burhel through my glasses at once – a sitting stalk – but unfortunately the fiat has gone forth that we are not to shoot even short of Phari. Thence we rode, seeing a herd of Tibetan gazelle grazing quite undisturbed under 200x from the track. Glorious view of Chomolhari: the weather wonderfully still & fine. Reached Phari about 1:30.

Spent whole p.m. until dark checking advance stores & wrangling with Dzong Pen.

5 4/24 *Phari*

Spent a busy day at stores all day – second party arrived about lunch time. Prolonged interview with Dzong Pen in p.m. – culminating in his capitulation.

6 4/24 *Phari*

Spent a.m. at stores, mess tent & a stroll with Shebbeare after birds. Dirty looking night.

Pl. 51 (L) Hinkey ambulans. [*Hingston*]

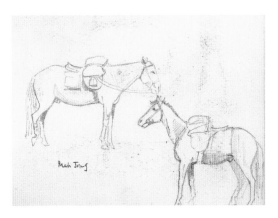

Pl. 52 (S) Mah Jong *and another horse.*

Pl. 53 (S) *Pencil studies of pack animals and a faint rear view of a Tibetan woman.*

7 4/24   *Chu ké*

Lovely morning. The anticipated transport trouble materialised with a vengeance, the Phari Dzongpen & Gembus obstructing in every conceivable way – Geoffrey & Shebbeare had an awful time, ultimately leaving at 2.30 p.m. Gyeljen proved himself very much worth his salt – Karma Paul the reverse.

Most of us started about 11.15. Somervell & I walked some 4 miles; I then rode in here another 4; found camp pitched in an E. & W. nullah very fairly sheltered from the wind, which was none too bad.

Our camp commands a particularly fine view of Chomolhari & its subsidiary ranges & glaciers; some of these are magnificent – one very fine rock & ice arête about the steepest & narrowest thing I have seen.

Darkness found it trying to snow – in accordance with the best traditions of this bit of country.

An absurdly short march of some 8 miles – stopping short of the Kya-rang (or Kyang) La which we cross tomorrow.

Kanza the head cook says this place is called La Dung : we all think Yak dung would be a more suitable name.

Thermometer max. 57°, min. at night 12°.

Pl. 54 (S) Don Q. The terror of the Sierras *and a man wearing a hat, thought to be Karma Paul.*

8 4/24   *Lung-Gye-dôk*

Marched about 9.30 after wearisome delay getting transport off. 12 or 13 miles march to our old 'Blizzard Camp' – of which I rode about 4 or 5 & walked the rest. It looked for a bit as if we were going to repeat our experience of 2 years ago (to a day) as it snowed on & off all day over both Chumulhari & Pau Hunri & we had some snow flurries.

But our luck held & though it was overcast – almost a monsoon sky – when we reached camp, it cleared towards evening & the wind was never bad (for Tibet).

The new mess tent is a great comfort on these occasions.

Rather worried to hear that the General (who is going round by Tuna & Dochen) has a chill & is staying an extra day at Tuna – Hazard has deviated to see him.

Our advance cooks are lost (since arrived).

Temperature in mess tent as I write at 4.30 is 40°.

Pl. 55 (L)   Chomolhari + Yak transport leaving Phari. 7 4/24.

Pl. 56 (S)   Chomolhari from Phari. 7 4/24.

9 4/24 *Lung dôk (16500)*

Got off at 10 – after an unpleasant night as it started to blow hard at 4 a.m. & with thermometer at about 4°F. produced a very cold morning. I walked to top of Donka La; went badly for 3 hours & then went A.1. – same old business of trying to walk & digest a meat meal at the same time.

We all lay off for an hour or more just below the pass out of wind & in sun; but arrived on top, the wind was so bad one could only ride short distances.

To make a long story short we at last descended off the 3rd pass about 5 p.m. & reached this place about 5.30.

Everyone played up A.1. & by 8 p.m. as I write everything is shipshape after a hardish day.

Thermometer in mess tent (which seems warm & pleasant) 28°F & outside 10°F.

Message received from Genl. to say he is better but is staying another day at Tuna to make certain.

As often before I am lost in admiration of the servants, porters &c – who walk 18 miles in bitter weather, get into camp & work like beavers to get their sahibs fed, clothed & housed.

The Donka La is undoubtedly & invariably a real stinker.

10 4/24 *Just below southernmost pass over Kampa (say 16500)*

Thermometer dropped to −2°F in night – but there was not much wind and personally I had a splendid night. Bright sunny morning – breakfast in the open.

Gyeljen having promised a very short march and the transport being in even worse form than ever, last yaks only left camp at 10.45 – Shebbeare and I doing rear guard, on foot.

About 5 miles on S. & I turned off to N. on ponies and lunched at Ang Zang Trak – out of wind and basking in the sun – the site of our camp after terrific march of 2 years ago yesterday.

Saw a pair of peregrine falcons – what I took for saker falcon last year no doubt: seen from underneath they look pure white.

Pl. 57 (S)   *Saddled horse.*

About 2.45 we rode to rejoin column whom we expected to find encamped 3 or 4 miles on. By dint of tracking found line of march and ultimately spotted the army going strong for the pass miles on.

We followed leisurely walking & riding – both our ponies inclined to be sore-footed.

Tibetan gazelle and kyang all round us – often 3 or 4 herds in sight in various directions and some within 300x.

Lovely p.m. and wind very light. It lulled entirely about 4.30 and then started to blow harder from S.W. – having blown all day from N.W. The first time I have noted this usual mountain procedure in Tibet.

Got into camp close under the pass at dark – an exposed and windy spot – I guess pretty low temperature.

Beetham is undoubtedly pretty ill with dysentry: Mallory is sick – tummy trouble of some sort. B. has had this for over a week and failed to report or take proper precautions early on – a pity. The rest of [*sic*] seem in good fettle: personally I am in rude health. We are all astonished at the exertions and hardships we went through last year covering this bit of country in 3 marches under really hard weather conditions.

Even doing it in 5 this year, marches seem long enough.

Somervell just reports temperature 25°F – which just shows how impossible it is to judge temperature in a wind – we all thought it real cold.

11 4/24   *Kampa Dzong*

Marched about 9: a glorious morning after a good night. Walked & rode quietly with J.G.B.

Ever since we decided that we must come via Donka La I have been looking forward to this descent into Kampa; it fully justified my expectations. The sky was cloudless blue, the country composed of every shade of pink, yellow & mauve – carbolic tooth powder, knife powder & wood ash mountains – the horizon encircled by row upon row of gleaming snow mtns. Everest & Makalu showing up in distance.

The Kampa – Tinkye plain looked very familiar.

Kampa was windless & the sun so hot that we had to unlace the end of the mess tent for lunch.

The Dzong Pen away; a capable looking Gembu promises us our transport for 14th & as we must give Mallory & Beetham at least 2 days to recuperate, this suits well enough & I hope will admit of the Genl joining up fit & well.

The 'troops' are remarkably fit & well – only one or two slight throats. We gave them a good feed of jam biscuits, rice and tsampha tonight & I have just left them cooking it all up with some of yesterday's meat into an enormous stew.

Sketched in p.m. [*Pl. 58*]. Glad to change clothes & spread out a bit. Some of our paraffin has leaked – I have allowed a big margin but I know it is all needed.

Pl. 58 (S)   Chomiomo from Kampa Dzong. 11 4/24. *On facing page:* The same an hour later. Snow and Chomiomo flushed pure pale pink with blue mauve shadows. All rest of background dark purply blue – pink ground [?]. Clouds on R rest of sky turning pale green.

| **12 4/24** | *Kampa Dzong* | **13 4/24** | *Kampa Dzong* |

Lovely morning – rather stormy in p.m. with a snow shower over Mendé. Quiet day in camp interviewing Gembu & odd jobs – Gembu can't produce transport until 15th now – so we have lost any start we had on 2 years ago.

In p.m. long discussion with Mallory & Somervell about plan of campaign. Then to Dzong with Shebbeare & Somervell – entertained by Gembu.

The Dzong really the most picturesque building in the world. Its squalor inside an extraordinary contrast to its fine exterior.

Invalids not too well.

Slack & uneventful day. In p.m. arrived the sad news that Genl has had to turn back from Tuna – for good, accompanied by Hingston.

So I now carry on.

14 4/24   *Kampa Dzong*

Spent whole day in camp & did a lot of work – hair cut, long wrangle with Gembu over price of transport (successful), wrote Times article, lunch, oxygen drill, discussion with Mallory over plan of campaign – conference with all climbers, wrote home, dinner & so to bed.

Invalids looking up.

Min. temp. 9°F.

Pl. 59 (L)   Kampa. 12 4/24. Hopeless. Everything drying and freezing.

15 4/24   *Linga*

Marched at 9 – Beetham being obviously on the mend, decided with Somervell to take him on at least as far as Linga.

Kampa Gembus played up well & transport got off in good time.

Rode to Mendé: thence with G.L.M & J.G.B. walked over top of hill to the 'Chung' village – a good 2 hours climb & descent & a view of Tinki lake.

Lunch at 'Chung' village & thence rode to within about a mile of here & walked in – pretty keen wind in our faces – must revert to windproof clothes.

A lot of cloud & – obviously – of wind on all the big mountains – not the sort of day to go high.

Min. temp. 11°F.

16 4/24   *Tinki*

Transport behaved excellently & all was away by 8.50.

Rode nearly all the way in order to tackle Dzongpens, &c.

Jolly march, hot sun, cold wind. Tinki lake looked lovely & was fuller of wild fowl than ever – red-headed & common pochard, wigeon, tufted duck, pintail, brahmini & bar-headed geese – were all spotted in one glance with binoculars.

Met by Shebbeare & party of Dzongpens &c. on horseback.

Spent p.m. sketching [*Pls 61–2*] & interviewing Dzongpens or rather <u>a</u> Dzongpen – as our old friend of 2 years ago is dead & his young relations reign in his stead. All out to help apparently – promise us to march 18ᵗʰ via Chiblung Chu.

Min. temp. 11°F.

17 4/24   *Tinki*

Spent day here – heavy & unpleasant day – presents to Gembus – visit to Dzongpen from 11 till 12.30 – interminable wrangle about price of transport & a meal: then oxygen drill. Then porters' drill with ice axes & rope: discussion with Mallory on the plan of campaign – another interview with Dzongpen (settled for rs 25 per animal for the 6 days) – round table conference on plan of campaign (settled at last) & then wrote home mail. Very strong cold wind from 11 a.m. onwards.

Pl. 60 (S)   *Pencil sketch of mountains across the Tibetan plain. Perhaps a distant view of Chomiomo.*

18 4/24   *Khenga*

Most of transport got off by 8.20, but 18 animals short & J.G.B. only got away with these by 12.30. In collecting them he got some of our own back off 3 Gembus, including the one who nearly gave us the slip at Gyanka Nampa with 16 yaks last year. He made these Gembus do a mile or two across country on foot at as 16 [*sic*] and then drive missing yaks back – leaving all prostrate & exhausted. He then reported to Dzong Pen, who entirely approved.

I rode flat but walked all uphill bits to Bahman Dopté pass – could not live with Mallory & Odell up last bit – disappointing, as I thought I was going to go A.1.

Ran down this side of pass & then rode in here – the village some 1½ miles S.W. of our last year's camp at To.

Perfectly lovely day – almost windless – lovely sunset effects coming down Chiblung Chu valley.

Again anxious about Beetham, who is not doing so well again. Gyeljen also none too well.

Pl. 61 (S)   Tinki. 16 4/24.

Pl. 62 (L)   Tinki Lake. 16 4/24. Impossible conditions, sun dust & wind.

19 4/24   *Chiblung*

A short march, say 10 miles – just as well as our last 18 yaks only got in at 10 p.m. last night.

Walked & rode alternately & tried to sketch [*Pl. 65*] – dull country at this season – missing the charm of green meadows & flowers we met in July 2 years ago.

Found Paul in camp – with English mail & letters from the General & Hingston. The rest of the porters got in in the p.m. after walking 28 miles from Tinki to catch us up.

The General definitely returns to Darjeeling – Hingston follows us as soon as possible.

Spent p.m. writing report to Everest Committee & Times article, with an hour's walk with Mallory between.

Weather singularly warm & windless (for Tibet). Min. temp. 13°F.

20 4/24   *Jikyop*

Marched about 9 after a delightful ¾ hour watching duck & wild fowl on a marsh through binoculars & telescope.

Then walked & rode alternately some 12 miles. Dullish march, weather wonderfully warm, but unsettled & later regular monsoon storms over Gyanka Range. Camped below village at actual entrance of gorge.

Spent p.m. writing Times article – fearful tripe – I only hope the British public will like it.

J. Macdonald went on to Trangsoe Chumbab in a.m. en route for Shekar in 2 days to make preliminary arrangements with Dzongpen.

After tea Mallory, Noel & I walked to sulphur springs just up main valley.

Min. temp. 14°F.

Pl. 63 (S)   *Unfinished sketch of Tibetan plain with mountains beyond.*

21 4/24   *Trangsoe Chumbab*

After breakfast crossed Chiblung Chu & some 500 or 600 ft. up hill opposite to see Everest. Got a perfect view tho' only just in time before usual cloud banner hid half mountain.

Sketched [*Pl. 66*] – then rode over sandhills to corner of Phung Chu – where we had lunch – warm & pleasant as last year.

Then Odell & I walked to camp – say 8 miles.

I make total march about 16 miles.

Stormy looking evening, considerable wind – again an overcast sky just like monsoon – what <u>is</u> the weather doing? Worked with Mallory at plan, & then tested Gurkha N.C.Os. at reading chits & did a little lamp signalling.

Oxygen wallahs very busy; the apparatus is failing all along the line – 33 cylinders out of 90 have leaked – almost every instrument leaks at every joint & tube. But for the mechanical genius of Odell & Irvine we should be hopelessly done. As it is we hope by radical reorganisation of the instruments to make up sufficient for the job.

After dinner allotted all climbers (tentatively) to parties. Min. temp. 13°F.

22 4/24   *Kyeshong*

Bad start as 200 transport animals were changed. We got off the last loads at 9.15; all donkeys bar 3 or 4 zohs – no yaks.

Rode to the 'ruined castle' partly along riverside – which much improved the march. After lunch walked to with[in] some 3 miles of Kyeshong giving Pasang Sherpa a lift on my horse.

Wind got up at the end, blustering but not really cold. Camped a mile above village in a pleasant bay – last transport only got in 6 p.m.

Spent evening working out cooly bundobust for glacier & other odd jobs.

Extraordinarily warm night 29°F.

Pl. 64 (S)   Pencil sketch of unidentified mountain range. The caption refers to Pl. 106.

Pl. 65 (S)   Nr Chiblung. 19 4/24.

Pl. 66 (L)   Mt Everest from above Jikyop (60 miles away). 21 4/24.

23 4/24   *Shekar Dzong*

Pleasant short march 10 or 12 miles – lunched at same old stream – where we found the bridge really well rebuilt – thanks probably to the efficient Shekar Dzong pen.

The colour on the variegated hills all round Shekar was wonderful – especially the crimson hill to North – veined by deep blue cloud shadows.

John Macdonald & 2 Dzong pens met us just outside Shekar – & we learnt that the latter all out to help.

Camped in the willow garden – surrounded by a wall, so that by mounting a sentry we secured welcome privacy from the mannerless inhabitants of Shekar.

After lunch Dzong pen came to tea & we fixed every point to our complete satisfaction in an hour – a striking contrast to our experience at Tinki. The Dzong pen here is capable, straightforward, direct & to the point – besides being a gentleman – a comfort to deal with.

Later finished report to Mt Everest Committee & Times article & began a letter to General.

One doesn't get much time to oneself these days!

24 4/24   *Shekar Dzong*

Climbed to top of dzong before breakfast with Shebbeare. Just got a view & sketch of Everest before clouds covered it.

After breakfast longish interview with Dzong Pen – who entertained us most hospitably & was very helpful & friendly.

Spent p.m. writing, checking stores &c. After tea Geoff & I about 1000 ft. uphill opposite – fine angry sunset.

Min. temp. 8°F.

25 4/24   *Pangle*

Walked with Mallory to within about a mile of camp – lunching at bridge. Lovely fine clear morning – but again spoilt itself in p.m. Some transport arrived late & Geoff only left Shekar at 1 p.m.

Several visited Gompa & presented head lama with a dud oxygen cylinder which they turned on & apparently rather alarmed the good men. 36 Tibetan coolies accompany us & were paid 3 days advance. Sprinkling of snow in night – temp. 15°F.

26 4/24   *Tashi Dzom*

Started 7.40 & walked whole march – reached Col in 1 ½ hours – magnificent view of all the great mountains. All sketched [*Pl. 67*], photo'd, examined mountain thro' binoculars & telescopes & sat about for an hour or more.

Then trundled down to gorge & sketched again for a bit [*Pl. 68*] & so in to camp in a delightful willow plantation.

Here we entertained the local Shegar or head man who rules over this valley & paved the way for future negotiations.

Spent the rest of the evening in financial calculations with Geoff – decided to get more money to Phari by Macdonald as a stand by.

Much colder, clearer & healthier day – everyone going well including Beetham.

Tashi Tundrup sick & fear pneumonia.

27 4/24   *Chödzong*

Getting near our goal. Overcast morning after warm night but cleared to a brilliant p.m. Dull march – one of the dullest.

Rode greater part. Lunched near mouth of Ding Chu & studied final arête of Everest with interest. Got into camp about 2 & spent p.m. writing for Times, seeing Tibetan coolies paid – a stroll with J.G.B. & some work on porters' parties for high camps.

Eyes worrying a bit – glare & wind I suppose, though I wear goggles.

Min. temp. 14°.

Pl. 67 (L)   Mt Everest from Pang La. 26 4/24.

Pl. 68 (L)   Below Pang La. 28 4/24 [*should be 26 4/24*].

| 28 4/24 | *Rongbuk Monastery* |

Uneventful march. Very cold wind – arrived at Rongbuk found the old Lama ill so he couldn't see us. Met our old friend the 'Dzong Pen' of Kharta & exchanged many compliments. Eyes still sore.

  Min. temp. 17°.

| 29 4/24 | *Base Camp* |

Walked up with Geoff – & all spent a very busy day arranging camp, sorting stores &c &c. Bitter cold, snow all p.m. – perhaps the coldest day I remember at Base Camp.

  Wind all night – Min. temp. 10°F, max. 49° – but the wind was the trouble.

| 30 4/24 | *Base Camp* |

Everyone again busy from morning till night – but such a change in weather – glorious sun all day & very little wind.

  Result of everyone's splendid work is that 150 loads have already gone to I (under Gurkha N.C.O.s) & 90 more go tomorrow.

  Geoff has done wonders organising this local transport. What would an Everest expedition do without him?

  Min. temp. 9°.

1 5/24   *Base Camp*

Sun hit my tent 7.30 exactly.

Greeted with the news that 52 of our local porters have deserted. However got off 75 loads from Base to I by using some of our own men. After an hour or so making up loads & typing lists of No III Camp stores – which have been repacked – Geoff, Shebbeare & I decided to go up to I – see how things were going & stay any further prospective rot among the local levies.

Started at 11 & by 12 had caught up the tail of the convoy: shepherded these into No I camp by 1.45.

Geoff there took hold, issued rice, promised another t2 a head for tomorrow's job – i.e. getting 100 loads into II. Found the Gurkha N.C.O.s in good heart & doing well. 78 loads had been put into II today.

One woman had carried a 3 year-old baby + her load to II & back.

I went badly – funny as we were only a few 100 ft. higher than the Pang La where I went well.

I was its old sunny self.

Weather on Everest fair – some slight snow showers but mostly fairly clear – not as good as y'day.

Returned to tea in 1 ½ hours.

Mallory meanwhile had been working at organization of the plan of campaign.

Noel's mules went very well to I.

Pl. 69 (L)   Two portrait heads of Geoff Bruce (wearing hat with hatband) and one of Odell.

2 5/24   *Base Camp*

Worked all a.m. at odds & ends – extra loads for high camps – clearing up camp, making out lists & log books for camps &c.

After lunch wrote & finished an old sketch until tea.

After tea walked up the gorge just above camp with Mallory & Somervell – going strong & wind A.1. – curious how one varies. Note received from Harke 'III Bhotia kuli le sab samn Camp II me achi tarah se parniche die' – written in Roman characters – a good days work well recorded!

Geoff paid off & fed all Bhotia labour but 10 retained for odd jobs & to accompany Hari Singh Surveyor.

So one more stage is completed satisfactorily & up to time beyond my expectations.

Mountain fairly clear in a.m. but one snow storm & a good deal of cloud in p.m. – windy – not a good day high up I guess, & a windy gusty night.

Min. temp. about 13°.

3 5/24   *Base Camp*

20 porters (our own) well loaded left for No. I about noon, & after an early tiffin Mallory, Irvine, Odell & Hazard left also.

Somervell, Beetham & I went with them as far as the highest point about a mile short of I: got there in an hour & 25 minutes, Mallory & Odell staying to whip in the tail of the porters who were not going too strong.

Back about 4 p.m.

Geoff reported some trouble with Shekar of Tashi Dzom who has written to Paul on the following matter.

At Rongbuk on 28th Geoff found 2 Tibetan porters had broken into a tin of Atta: as we had had several similar thefts (including ²/₃ of the porters' high altitude boots) he asked the Chunji La to make an example of these men & the Chunji duly flogged them.

Now the Shekar writes objecting & withholding supplies of champha & butter until he gets an answer.

Paul goes down tomorrow with a letter from me to deal with him. I don't anticipate international complications.

Cold stormy day – mountain very windy & cloudy all a.m. & obscured by snow scurries all p.m. – a bad day high up.

Min. temp. 13° - very gusty cold night.

Pl. 70 (L)
Porter carrying load.

4 5/24   *Base Camp*

Got off 20 porters under Umar at 10.45 – lightly loaded. These to go to II tomorrow & be based there, carrying to III, until redistributed on 8th.

Paul left 8.30 for Tashi Dzom. Easy morning for 5 of us in camp. Cold wind, hot sun in a.m. Very strong wind & a lot of cloud on mountain.

Walked up western trough of main glacier with Geoff & Shebbeare before tea.

Min. temp. 15°. Snowed all the latter half of night.

5 5/24   *Base Camp*

In a.m. got off Hari Singh (surveyor) with written instructions – then wrote letters & underwent Hingston's tests administered by Somervell.

After lunch Geoff & I for about an hour up the E. trough of main glacier & back to tea. Packed for tomorrow.

Slack quiet day.

Starting by being very cloudy, cold & evidently very strong wind up high. Gradually cleared & towards evening Everest was clear. A bad day high up but looks like finer weather.

Min. temp. −4°F, but no wind & warm in bed.

6 5/24   *No. I Camp*

In a.m. finished packing kit & got off 5 porters by 11.30. Lunched early & left with Somervell at 1.15 – Beetham ahead photographing. Walked very quietly & arrived No. I 3.45.

Checked stores after tea.

Lovely day. Mountain looked good all day though wind increased towards evening – No. I its old sunny self.

Pitched 2 Whymper Tents, left 2 beddings, & so completed establishment of camp.

Sun left tent 6 p.m. Min. temp. 5°F.

7 5/24   *No. II Camp*

After mixing oil & odd jobs left at 10.10; went v. badly (? not enough breakfast) & got into No. II 1.45.

Found things far from well. No. 2 party had been up to III & back 2 days running & No. 1 party instead of arriving this a.m. & going back filtered down this evening to spend the night here.

This entailed 40 odd men sleeping in a camp designed for 20 &, in addition, these new arrivals appeared dead beat & complained of great privations & hardships – only tzampha to eat, no extra blankets, frostbitten feet, &c &c.

I did every thing in my power to fix these men up. Issued tents, blankets, food, fuel from high camp dumps.

Our shortage of transport officers & of people who can really talk the language is going to be a very severe handicap.

Min. temp. –1°.

8 5/24   *No. II Camp*

Mallory arrived to bk'fast from III & explained much of what had happened – they had had –21½°F the first night & very low temperatures the next night.

Owing to the parties from No. II dumping at a point short of III no rations or extra blankets had reached them: they had lived on tsampha & been perished with cold.

We made new plans. Somervell went off with 14 of No. 2 party porters at 10.45 – unloaded as far as dump, thence to hump all loads essential for their own comfort first up to III.

Geoff arrived with 7 reserves (from I) & proceeded to get things happier more suo.

Mallory & I stayed at II & spent the day reorganizing plans. Sent for Hazard to come down to Base, take over from Shebbeare & send latter up to II. Rearranged loads &c, &c all day.

Sun left camp 4.45 – turned into flea bags until dinner – Min. temp.

This day Odell & Hazard made an attempt at N. Col; only got up about ½ distance & dumped pitons, rope &c near old bergschrund. [*This sentence added later.*]

9 5/25 [sic]   *No. III Camp*

Got off 26 porters; 18 of 1st party to go to dump, 8 to go right through to III – Mallory accompanying. Geoff & I left at 9.15 – route over glacier considerably changed. Glacier surface composed of absolutely smooth ice varied by depressions full of powdery snow. Crampons a great help.

Met Hazard coming down – beard encrusted with ice. [*This sentence added later.*]

Followed the trough for quite a distance – wonderful scenery – surrounded by fantastic blue pinnacles like a Drury Lane transformation scene – G & I went well & easily & arrived at No. III about 2. Snowed all day, gradually getting heavier & assuming proportions of a regular blizzard.

After warming up in tent for a bit turned out & went round camp – very miserable affair – Odell & Irvine got roarer cooker going, we turned out porters with cooking pots & got some hot food going.

Dined & turned in about 5.30 or 6 p.m. Blizzard increased & all night snow drifted into our tent, so that by morning whole tent was an inch or two deep in drifted snow covering any spare clothing, &c. lying about – pretty miserable.

Min. temp. +10°F.

10 5/24   *No. III Camp*

Decided Mallory & Irvine had better go down to II – prepared them for possibility of evacuating altogether next day if bad weather continued. Sent to Noel & Beetham not to come up.

Somervell & I with about 16 porters descended to dump & brought up loads, carrying ½ loads ourselves, in teeth of horrible wind. Snow had stopped falling but was being blown along surface of glacier producing the same effect as blizzard.

Returned to camp, busied ourselves with getting enough primus stoves going for porters' cooking.

Blew harder as evening came on in tremendous gusts from any & every direction.

Spent a restless night – tent shaken like a rat by a terrier – snow piling in through flaps of door – Decided evacuate next day. [*These words probably added later*] Min. temp. −7°.

11 5/24   *No. II Camp*

Temp. at 9 a.m. still −1°F & blowing great guns. Decided to evacuate whole line – North Col route palpably unsafe for some days – all porters more or less done in – sahibs not improving much.

Geoff & I turned out & did all we could to strike camp & check stores &c. I wrote a list of stores for a minute or two & had then to go to ground in tent for 10 minutes to restore feeling to my hands. Geoff was splendid & got a wonderful move on the men who were quite numb & useless. Result all tents struck & packed in bags, & with all stores &c. made into a dump, & party got off in good shape.

Conditions soon improved down glacier – that is as regards wind; glacier lassitude supervened instead & it was very hard work down hill.

Met a party from II making for dump – my chit sent off first thing having arrived too late to stop them.

Then met Irvine with news that Tam Ding had broken leg – & been carried back to II by Dorjay Pasang.

Pl. 71 (loose sheet)   Geoff Bruce addressing the porters in the blizzard on the East Rongbuk glacier.

Arrived at II decided that G. & I would remain along with No.1. party who had carried loads to dump – our No. 2 party from III to go down to I with Somervell & Odell. Mallory, Beetham, Noel & Irvine to Base.

Spent shocking night much like that of 20 5/22 – & found in a.m. that I had no mattress – Geoff having (unwittingly) collected both – Tam Ding groaning in next tent.

Min. temp. 0°F.

12 5/24	*Base Camp*

Got things packed & everyone off by about 9.45 – warm & sunny in camp. Tam Ding on sick-man carrier with 5 men in attendance, carrying in turn – travelled pretty comfortably.

Arrived at I, found Somervell in attendance on Man Bahadur very badly frostbitten feet, Sangloo bad bronchitis, & L Nk Shamshar insensible & paralysed – apparently haemorrhage or clot of blood to brain due perhaps to frostbitten fingers.

Decided last must be left absolutely quiet, so left him with Herke & Nursang & went on ourselves to Base Camp with other sick men – arriving about 1.30.

Met Hingston & Hazard en route – former arrived the day before.

Bathed & changed clothes in p.m. & had a perfectly gorgeous night after some colossal meals.

Min. temp. 2°.

Pl. 72 (loose sheet)   Geoff Bruce addressing the porters in the blizzard on the East Rongbuk glacier (second version).

13 5/24   *Base Camp*

Sent off Paul to find if Rongbuk Lama will bless whole entertainment. He will on 15th.

Spent whole a.m. writing communiqué for Times & p.m. till tea writing up this diary – After tea Geoff & I reallotting porters' parties.

Hingston & Geoff to No. I; found Shamshar still unconscious & arranged his return here on an improvised stretcher.

In p.m. all hands packing fresh stores for high camps.

Hingston working hard at sick. Man Bahadur must lose part or all of both feet – Tam Ding doing very well.

Sangloo & one other bronchitis case not too gaudy.

Rest of porters immensely restored already. All sahibs fit & well. It is very bitter to have to give it best – even pro tem, but

'No game was ever worth a rap for a rational man to play

'Into which no accident, no mishap could possibly find its way.'

Poor Shamshar died on the moraine shelf ½ mile from camp as he was being carried in.

14 5/24   *Base Camp*

Shamshar buried about 10 a.m. A day of reorganisation. Collected spare sahibs' kit for porters – held kit inspection, trained cooks & new porter N.C.Os with primus – wrote home, checked Times communiqué; after tea walked with Geoff for an hour.

Snow in p.m. – mountain looked quite damnable all day – very cold wind.

Mallory, Somervell & Odell went for a climb.

15 5/24   *Base Camp*

All (bar Somervell, Beetham & 3 or 4 sick) left for Rongbuk Monastery about 9.45 – arrived there, we were ushered into a sort of anteroom & for an hour & a half killed time, drank tea & ate macaroni & mutton.

Then we went upstairs to an open air court at one side of which the head lama was enthroned in state on an iron bedstead – beautifully dressed & surrounded by satellite lamas – a most impressive man – courteous, dignified & with a curiously plastic, humorous face full of character.

We all stepped up in turn & were blessed – he touching our heads with a silver prayer wheel. Then all the men came up, making deep obeisance, & going through the same ceremony.

From time to time the old gentleman sent me a friendly message by Paul. He made a short address & led a chant of 'hum mani pedmi hum' in a splendid bass voice.

Then I gave our presents: we had a cup of rice apiece – received a present of eggs & salt from him, & finally left after further exchange of courtesies.

We all came back much impressed by what we take to be a genuine saintly man.

The 'troops' much heartened. As Paul characteristically puts it 'The men are all very thankful to the Babu & the Colonel Sahib for arranging this ceremony'.

16 5/24   *Base Camp*

As a result no doubt of y'days function this morning broke as weather should here at this season – brilliantly clear, quite still, no cloud – & all day, tho' the usual wind got up, no clouds bar the usual streamer on Everest, appeared – down the valley, sky cloudless brilliant blue. At once decided to resume attack tomorrow – got off some 16 loads to I under Gurkha N.C.Os.

Spent morning in conference & sub-committees on details of organization, fuel, loads, &c. & p.m. in getting same tabulated – (so that in future all sahibs may take an intelligent interest in the organization) & in writing letters.

Head of Burhel
feeding
within 25×
of Base Camp
Cook House
16/5/24.

Pl. 73 (L)   Herd of Burhel feeding within 25x of Base Camp cookhouse. 16 5/24.

Shall be glad to be off again tomorrow – the Base Camp is only possible when the job is <u>done</u>.

By the way herd of burhel grazed within 25x of cookhouse this a.m., whole camp looking at them. They absolutely unconcerned.

17 5/24   *No. I*

The day we hoped to bring off our 1st attempt & the most perfect day in these parts I ever saw – almost cloudless, wonderfully warm, almost windless – Dis aliter visum.

Anyway we start under good auspices.

Got off 26 porters about 11.30: had an early lunch & got away ourselves about 1.50 – Mallory & I reached No. I in 2 hours going very comfortably – Shebbeare close behind – Somervell & Odell followed leisurely.

No. I a sun trap – all lay & basked until about 6 p.m.

Porters all seem well & happy.

18 5/24   *No. II*

Mallory, Somervell, Odell, Shebbeare & I left No. I 9.30 & after halting say 25 min. en route arrived No. II 12 noon – feeling good.

Porters took a bit longer but came along well.

After lunch we all distributed to various jobs:-

Shebbeare – comfort of camp &c.

Somervell & I arranged loads for next 3 days.

Mallory started in on loads from III to IV, checked No. II Camp stores, &c.

Odell did oil, primus, Meta &c.

Weather remained fine & clear to N. but there was a good deal of cloud off Everest coming near this camp & it was colder than at I y'day.

On the whole it looks good enough but it's anxious work.

Min. temp. 0°F.

19 5/24   *No. III*

After 26 porters despatched with Mallory & Odell to dump, Shebbeare & I went into all loads again for next 2 days & then followed.

Caught up at dump & made lists &c. Then went on, I carrying a light Meade tent, 2 coolies' blankets & an oxygen carrier – went well – some affected by sun & glacier lassitude in trough – particularly Somervell & Shebbeare.

Arrived at III found Mallory doing good work getting up tents &c. Spent busy p.m. squaring up & taking stock.

Fine but cold at III – Good night.

20 5/24   *No. III*

Neither Somervell nor Mallory fit in night, so I started to take place of either to N. Col.

At foot of slopes it became evident that S not fit – so turned him unwillingly home.

Mallory, Odell & I went on with Llakpa Tsering to carry rope & pitons – O. made mistake of trying to carry pitons & got rather exhausted in consequence.

Worked out a new route – designed to be safe from avalanches – route went well bar an ice chimney 150 feet high, up which we shall have great difficulty in carrying loads.

To make long story short Mallory & I divided the arduous work of cutting & kicking steps right up to old camp site where we arrived 2.30 p.m. – a fine day's work & one I am glad to have taken part in.

Arrived there M. & O. reconnoitred as far as N. Col & established possible route – O. playing up well – while I drove a couple of pitons for fixed ropes.

We actually only fixed one rope (in chimney) but decided position of all others & dumped rope & pitons handy.

Left top 3.45: Mallory perhaps mistakenly decided to take alternative (1922) route & rattled on ahead – he & I unroped – Odell roping Llakpa.

On a certain steep icy bit I slipped & went some way but turned over, got head of axe under me & pulled up – no damage, but nasty.

In same place Llakpa also slipped & tho' Odell held, the knot (a reef) tied by L. came undone & L. only pulled up by chance.

We then had to come v. slow, L. obviously badly shaken, I had once to cut steps up 50' or so to meet him.

Meanwhile Mallory was in bad trouble all alone; he fell 15 feet down a crevasse & twice more into crevasses.

Finally we all reached glacier about 5.30 p.m. considerably the worse for wear & M. obviously exhausted & shaken. (He was quite unfit to start in a.m. but has the most wonderful heart.)

I came in surprisingly fresh & strong for me.

Indifferent night – head too full of the very apparent difficulties & dangers of the whole business.

Overcast & warm night, light snow in early a.m. – don't like look of weather much – pray Heaven its not the beginning of monsoon as no power on earth can make parts of N. Col route safe under monsoon conditions.

Pl. 74 (L)   T.H.S[*omervell*] coming back from N. Col. 1924 [*probably 21 5/24*].

21 5/24   *No. III*

Morning broke warm with a lot of cloud about, light dry snow kept falling at intervals, but sun coming through & we anticipated no trouble.

Got 12 porters off about 8.30 for N. Col with Somervell, Hazard & Irvine in attendance – S. & Ir. to return, H. to remain & hand over to J.G.B. & Odell tomorrow.

Up till 1 p.m. N. Col slopes remained clear but no sign of party along big crevasse.

THS coming back from N Col, 1924 (Somervell)

Soon after it began to snow in earnest & nothing more could be seen. As I write at 6.20 p.m. S. & I. have not returned & it is still snowing – very soft wet snow – I am anxious.

G. Bruce, with 19 porters arrived from II about 2 p.m. having cleared II & practically all dumps.

I spent whole day on loads, stores & bundobust & am now so sick of it all that I can hardly bear to think any more about it.

6.35 p.m. S. & I. just in – party reached camp – so all is more or less well. It has been snowing since lunch time.

22 5/24   No. III

A perfectly bl-dy day. Snowed all night & all day until about 3 p.m.

Impossible to do anything – Hazard & party of course marooned at IV. Couldn't even see them till 4 p.m.

Lay in flea bags nearly all day reading home mail & papers – feet like stones all day – thoroughly unhappy about things in general as this snow looks very like a foretaste of the monsoon & anyway we lose a precious day.

Turned bitter cold at night –24°F. – the coldest ever recorded in these parts. Many of the party hardly slept & most are below par. I keep well thank God.

23 5/24   No. III

J.G.B., Odell & 17 porters left for No. IV by about 9.30.

Decided to send them as last night saw 2 of Hazard's party descend the worst part of traverse just below IV Camp.

J.G.B. as usual very much all there.

Cloudless brilliant day: no wind but very keen cold air.

Porters left in very fair form.

Then spent morning collecting missing tents, checking stores, writing down to No. II, sending to dump for more oxygen, &c &c. Started snowing steadily about 1 p.m.

About 3 p.m. Odell, Geoff & porters returned: they had got about half way up, encountered deepish snow in dangerous condition & decided it unsafe to go on – so dumped stores.

They saw Hazard & part or all of his party descending above them.

About 5 p.m. Hazard arrived, but with only 8 of his party. He had for some unknown reason told Phu to remain, 3 more men were either sick or funked the traverse at top & turned back, remaining there with Phu.

Here was a pleasant situation! especially as it transpired that a load of mixed porters' rations had been dropped off the steep traverse at the top & the marooned party had nothing but tzampha to eat – besides sahibs' rations.

Meanwhile it was snowing harder – regular monsoon snow.

Decided once more to evacuate – route to N. Col palpably unsafe – all porters have had a bellyful – Irvine & Odell both sick men – Mallory bad cough – Somervell miles below par – Hazard O.K. but had enough pro. tem.

Finally conditions all point to a real monsoon current & while this lasts cannot risk marooning parties on N. Col – neither can we continue burning fuel at No. III.

Meanwhile Mallory, Somervell & I are to go & rescue maroons tomorrow – a pleasant job!

24 5/24   No. III

Heavy soft snow until midnight followed by min temp. of –16°.

I slept little thinking of most unpleasant job ahead of us (Mallory & I afterwards agreed that we both put our chances of being avalanched at about 2 to 1 on) & of horrible situation if we failed to rescue these men – chances of some being too frostbitten to come down with us & so on.

It is not all gain commanding an expedition of this sort.

Mallory, Somervell & I got off about 7.30 – bitter cold morning, air temp still –2°F after sun had been up 2 hours.

Geoff meanwhile evacuated camp – bar Odell, Noel & some 5 or 6 men to help down any casualties next day.

We three crept over glacier – going so badly that my hopes of our accomplishing our job were very low – the fact is that one cannot stand a succession of these cold nights & snowy days without sad deterioration.

However once off the glacier we improved – led in turn – Mallory to ½ way up first slopes, S. to first crevasse & I then to foot of chimney. Up this Mallory led (50 min for 3 of us to get up) & on to corner of big crevasse – I then led what we regarded as first dangerous bit up to serac, M. & S. roping me from below & then I them from above.

At serac Mallory & I roped Somervell across the dangerous traverse at the top.

He had a horrible job & did magnificently. Phu & Namgya slipped just as they reached him & the rope (which was 10x too short) & for a moment it looked like tragedy – but they pulled up on their own after 10x or so – S. as cool as a cucumber – passed everyone across on the rope successfully, giving a first-rate exhibition on a d–d nasty place.

Namgya had badly frostbitten fingers – & feet less so – Uchung feet slightly frostbitten – Phu fingers slightly.

This party we started to get down at 4.30. In the chimney I held Namgya's full weight at least 4 times. I had crampons & so could do what I couldn't otherwise have done.

The 4 men behaved splendidly & needless to say S. & M. were perfect towers of strength.

We reached glacier about 7 p.m. & camp 8 p.m. after being met by Odell & Noel with soup en route.

Both awfully good to us here & later in camp.

A gruelling day – powdery snow 6" to 12" deep all day – once up to our waists – but a triumphant success in the end – cold feet all day & all next night. [*Last four words added later.*]

Min. temp. –13 °F.

25 5/24   No. II

I got v. little sleep as my feet had suffered a bit during day & wouldn't really warm up.

Got camp evacuated about 10.45 – usual procession of sick men.

Walked down to II with Mallory – glacier all snowy again.

Arrived at II sent Mallory, Shebbeare, Geoff & Irvine down to I – Beetham, Somervell, Odell & I remain.

Snow showers from N.E.: weather looks very hopeless. Believe a monsoon current sweeps up Phung Chu & approaches us from that direction. This weather is cruel; with 2 years ago's weather I believe we should now be homeward bound with the mountain defeated.

Camp II is not exactly an earthly paradise but it is a peach to III this year.

I haven't seen the latter clear of snow or had a meal in the open there this year.

26 5/24   Camp I

Got off the miserable little convoy of sick & wounded – & then walked down to I with Somervell; both going fairly strong.

After lunch at I held council of war – results indecisive & propose repeat next day with possible addition of Odell & Hazard from II & Base.

Good night – min. temp. +10°F.

I maintaining its character – weather fine – a lot of cloud in p.m. but dissipated at sunset.

27 5/24   Camp I

Wrote communiqué & letters all day – except for council of war – which decided to scrap oxygen & go for series of small attempts of 2 members each. S. & I to form second – M. & J.G.B. first.

Glorious day; sun too hot at times.

28 5/24   Camp I

Much same as y'day – everyone undoubtedly benefiting by rest. Irvine left for No. II to make rope ladder.

Redistributed porters, putting the 'Tigers' at Camp II.

29 5/24   Camp II

On Ascension day we start again.

34 porters left for II & after the usual settling up Mallory, Somervell, Geoff & I left at 10.55 & walked through to No. II in 1 hr. 55 min. going comfortably but without stopping – getting in for lunch.

Spent lazy p.m.

Odell & Irvine finished construction of a very good rope ladder made of Alpine rope & tent pegs for the chimney.

Perfect weather.

30 5/24   Camp III

Got off 13 porters – nearly all at last properly equipped with ice axes, crampons & carriers.

Geoff & I left at 10.10 – Mallory, Noel, Beetham & Somervell about same; crossed trough & followed bosom of glacier – v. good going on crampons.

Went slowly but comfortably until last mile when snow & absence of wind produced some glacier lassitude – got in by 12.30, 2 hrs 20 min. including a 20 min. halt & smoke at upper dump.

No. III more like its old self – snow almost all gone – & a very little water to be found by digging in glacier.

Lunched in open.

In p.m. Mallory, Geoff & I had the usual discussion about loads &c. & got everything thrashed out.

Porters are going to be short & shall have to rope in at least 2 more from II tomorrow.

Clouded over towards sunset but cleared again.

Everyone slept like a top – some for first time for some days.

Min. temp. +5°F.

31 5/24   Camp III

Fine sunny morning tho' some wind – high cirrus clouds – top of Everest clear.

Geoff, Mallory, Odell & Irvine with 9 porters got off about 8.45 – good luck to 2 former. Noel with them, S. & B. to Rapiu La.

Give us 3 more days of fine weather & we may do it yet.

Pl. 75 (L)   Profile view of climber, probably Geoff Bruce.

Pl. 76 (L) Pencil studies of climber, probably Geoff Bruce.

N.B. All the following is written from memory 8 days later at No. III.

1 6/24	No. IV

Somervell & I with 6 porters left for N. Col – 9.15 – weather still perfect.

Got up in good time – Camp IV 3 p.m. – doing ice chimney in 1 hr. 10 min (thanks to rope ladder). Irvine & Odell cooked for & waited on us. Arranged loads &c in p.m.

2 6/24	*Camp V*

S. & I with our 6 porters got off about 6.30 in fine weather – crossing the Col – out of shelter & in shade we struck bitter wind.

Proceeded up ridge & ½ way to V met Dorjay Pasang with note – & soon after Mallory, Geoff & 3 porters coming down. It appears they had had a bad time with wind the day before & porters – headed by D.P. – refused to go on. Attempt had to be abandoned.

We continued & reached No. V at about 1 p.m. & thence sent back 2 porters, keeping 4.

Spent the usual miserable p.m. cooking & filling Thermos flasks. Failed to infuse any life into 4 porters whose condition was not improved by a fall of stones; these cut Lobsang Tashi's head & Semchumbi's knee.

Spent a fair night.

3 6/24	*Camp VI*

Turned out early – 5 a.m. – & tried to get porters to cook themselves breakfast: cooked our own.

Used every persuasive power I have on porters & finally got Nuboo Yishay, Llakpa Chedi & Semchumbi (all honour to him, knee & all) to agree to come on: Lobsang Tashi, sick, went home by himself to IV. Got off about 9.

Uneventful climb – less wind – pitched Camp VI (some 100 or 200 ft. higher than highest spot we reached in 1922) at about 1.30 (Somervell disagrees about these times).

T.H.S. with bad cough & sore throat going badly & couldn't keep up.

I reconnoitred on – next day's route a bit, while porters pitched tent in a cleft of rock on the ridge.

Sent porters down about 2.30.

Usual routine – continually fetching & boiling snow – eating loathsome meals & filling Thermos.

Spent the best night since I left Camp I.

4 6/24 — Camp IV

T.H.S. pretty seedy with throat. After bk'fast (for wh. we had to boil snow as one Thermos had leaked in my bed at night) we got off at 6.40 – made a line diagonally across yellow strata towards foot of pyramid.

Had to wait continually for T.H.S. – weather fine, more or less windless but so cold that one shivered continually despite layers of windproof clothing, & when in sun.

Near big couloir at noon T.H.S. had to give it best – I went on alone but almost at once encountered v. bad going – no footholds on sloping slabs & much snow lying soft & powdery on these slabs. Crossed big couloir but only made perhaps 500x & 100 feet on T.H.S. & at 1 p.m. turned back (28,128 ft) [*height added later*]. Joined him by 2 p.m. when we turned home. He lost his ice axe.

After some scrambling made VI, picked up sacks & so down ridge – horrid sliding scree – T.H.S. going v. slow sick man. Got to pt. level with V at sunset & onto snow at dusk. Here I glissaded but T.H.S. dropped ½ hour behind & I had to sit & wait. Eventually hailed IV, gave T.H.S. my axe & took his tent pole.

Mallory, Odell & Irvine with oxygen met us 200 ft above Col & escorted us back to IV (9.30), fed & looked after us royally.

At 10 p.m. I developed real bad go of snow blindness, v. painful, stone blind.

5 6/24 — No. IV

Stone blind all day & in considerable pain. Helped G.L.M. to complete plan for his & Irvine's attempt with oxygen next day. Talked to porters a little & so on.

T.H.S. left for III in p.m. Hazard came up.

6 6/24 — No. III

Still stone blind after a night of pain.

Mallory, Irvine & party got off about 7.30.

About 10.30 Hingston, Nima Tundoo & Chutin arrived to look after me.

Decided to go down, stone blind as I was – Hazard escorted & roped me as far as bottom of chimney & the other three the rest of the way – a notable performance on the part of Hingston.

Was carried into III from edge of glacier down the moraine on one-man-carrier & got in about 4.30 p.m.

7 6/24 — No. III

Spent a night without pain & next a.m. could begin to see a little after some 60 hours of complete blindness.

Decided must stay III until Mallory's attempt was finished.

Lay up all day. Camp extraordinarily warm.

In p.m. got a message by porter from Mallory at No. VI saying both he & Irvine had got along well with oxygen, proposed to start early next day & telling Noel to look out for them at foot of final pyramid about 8 a.m.

Hingston reports my heart dilated – pretty rotten.

8 6/24 — No. III

Eyes much better. Spent whole a.m. dictating Times communiqué to J.G.B. – weather changing: much soft cloud about largely obscuring summit of mtn. which was peppered with snow by p.m. Camp very warm again.

Noel could see nothing of climbers all day though he kept his eye glued to telescope.

In p.m. interest & anxiety increased. Ultimately Odell (who had slept at V the night before & gone on to VI geologising) was spotted towards sunset descending via V to IV.

No sign of others – nor any signal from IV with whom we had arranged signal to indicate success or failure of climbers.

9 6/24   *No. III*

Rough, cold day – very stormy high wind on mountn, mostly from N. but conflicting with other currents from E.

Spent whole day in camp with people watching mtn, Nos IV & V camps &c all day.

Anxiety momentarily increased.

By 11.10 decided disaster probable – sent letter of instructions to Odell & Hazard

(i)   Reconnoitre to VI if could do it within 24 hours & if they considered it possible climbers could be there unseen

(ii) Clear Camp IV by 4 p.m. tomorrow or earlier if dangerous snow begins

(iii) <u>Principle</u> to risk no more lives in trying to retrieve the inevitable

(iv) Within these limits act on discretion

Soon after this letter left, saw what we took to be Odell & 2 porters leave IV, apparently for V tho' could not spot them arriving there. Shebbeare & some 30 porters arrived from II at 11 a.m. to evacuate camp; but could do little under existing circs so sent them back with 4 or 5 loads at 2.30 p.m. to return on 11th.

Of all the truly miserable days I have spent at III this by far the worst. By now it appears almost inevitable that disaster has overtaken poor gallant Mallory & Irvine – 10 to 1 they have 'fallen off' high up.

6 p.m. my messengers to IV returned.

Hazard has 'sent Odell' (God knows why he didn't go himself) to V – one porter to go on to VI if possible tonight – Odell not to go beyond VI.

Mallory & Irvine were last seen by Odell from VI (12.50 p.m. on 8th) 'going strong' about the final step before the pyramid – this appears to confirm my fears as above.

10 6/24   *No. II*

Very similar day to y'day – straining eyes on the mountain all day.

About 2 p.m. received Hazard's signal to the effect that Odell (who had gone on alone to VI) had found no trace there.

Waited until 2 porters came down at 5 p.m. with confirmatory letter & at 5.30 saw Odell reach IV.

Geoff & I then left for II – started strong but finished groggy; we are neither up to much & I am, as in '22, foundered in the feet.

11 6/24   *Base Camp*

Came through by lunch – with ½ hour at I.

Base Camp smiling & almost green – what a change!

12, 13, 14 6/24   *Base Camp*

Writing wires, letters, communiqués &c. all day & every day. Odell, Hazard, Hingston all arrived O.K. Shebbeare evacuated all camps. Started on monument &c. &c.

15 6/24   *Rongbuk Monastery*

Monument a great success – much photographed.

Marched about noon & spent quiet p.m. at Rongbuk writing home letters & inspecting ponies &c.

16 6/24   *Chogorong*

Very pleasant, peaceful march after saying good bye to Noel & our Shekar contingent. Camped at a dhokpa high up above Chöbu – a delightful situation & wonderful view of Everest – sketched [*Pls 80–1*] & was more peaceful & content than for a long time.

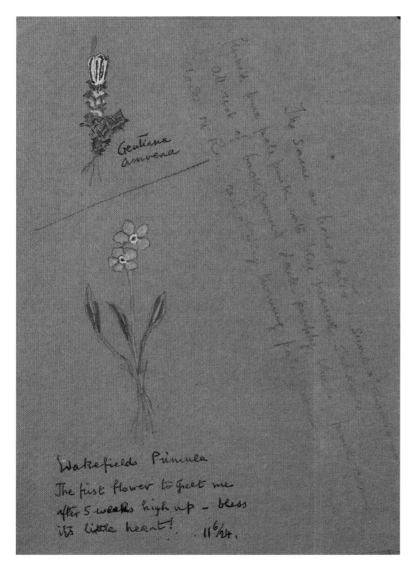

ABOVE: Pl. 78 (L)
The Memorial Cairn
with Everest behind.
Probably 14–15 6/24.

Pl. 77 (S)   Gentiana amoena. Wakefield's Primula. The first flower to greet me after
5 weeks high up – bless its little heart! 11 6/24. *[Also pencil annotations for sketch*
of Chomiomo on opposite page, Pl. 58.]

RIGHT: Pl. 79 (L)   The Memorial Cairn. Probably 14–15 6/24.

17 6/24   *Sichu*

Geoff & I walked over Ponding La (not Lamna La as marked on map) – I
went quite well despite my feet which are affected in exactly same way as
2 years ago. Fine view northwards over Tibet from Ponding La – Te'n Ri
a magnificent double mountain 60 miles N.N.W. attracted much interest
& attention.

Rode down over desolate plains (kyang & burhel country) to a
dhokpa, where we camped. Monsoon boiling up to S. & evidently weak
so far. Colder.

18 6/24   *Kyabrak (not Kyetrak as marked on map)*

Long & dull march up interminable stony valley – scenery getting grand
as we reached the head. Cho Rapsang a magnificent glacier-covered
mountain blocking head of valley – but obscured in rain & snow clouds
all p.m.

Kyabrak a tiny, squalid village situated on a great alluvial fan about a
mile or so short of snout of great glacier.

Pl. 81 (S)   Everest from Chogorong. 15 6/24.

Pl. 80 (L)   Chogorong. 15 6/24. *[unfinished]*

The road over to Sola Kombu via the Nangba La (<u>not</u> the Kombu La) follows the trough on E. bank of glacier – important trade route to which the village of K. owes its existence.

Monsoon stronger today – light rain as far as our camp.

19 6/24   *Tazang*

Followed trough on L. of glacier for a bit & then turned R. handed towards Pusi La. As one mounted got a grand & wild view, Cho Uyo & its satellites towering over the glacier which ran like a great high road over into Nepal. Pusi La about 17000 – I walked it with great ease – heart must be pretty well O.K. tho' feet still sore.

Looked over into a wild & desolate valley with very fine cliffs. Descending passed close under some very fine snow peaks & great glaciers & ice fields – obviously much more snowfall here. Country already much greener & many new flowers & birds.

Followed valley for miles getting steadily narrower & deeper – eventually became a deep gorge with towering cliffs on either side – something like Gautza gorge.

2 sorts of dwarf rhododendron but disappointingly scarce – vegetation steadily increased – juniper, cypress, willow & many other shrubs – near Tazang met golden birches – but so far no conifers.

Reached Tazang 4 p.m. & transport all in 2½ hours later.

Delightful soft mild air – what a change! Clouded up pretty thoroughly in p.m. but no rain.

A glorious day! The most conspicuous feature was the sweet aromatic smell of shrubs & vegetation which met us near the pass long before we saw anything green.

Pl. 82 (L)   Cho Rapsang + Midsummer at Kyabrak. 11 7/24. [*The date is wrong. The diary records passing Cho Rapsang on 18–19 6/24 and 3 7/24.*]

20 6/24   *Between Tazang & Tropdé*

Marched all down gorge – disappointing – gorge hardly varied & never opened out – vegetation of course increased but never approached Kama valley form – rhododendrons conspicuous by their absence – primulas &c few & far between. No view at all – no real forest. Birches, mountain ashes &c. on opposite side of river – on ours only stunted cypresses & a few larches & pines. No big trees at all so far.

Got to camping ground about 2.30 & found mess tent &c had gone on – yaks only got in between 4 & 5.

21 6/24   *Tropdé*

Walked all the way: feet fairly sore. Gorge got more interesting, flowers &c improved – wonderful show of irises & roses of two colours – but never approached Kama valley form.

    Mist on mountains all day & never got a view.

    Selected only possible camp just above Tropdé – & not too bad at that – on a tributary torrent under a big cliff – irises & roses all round our tents.

    Wet p.m. – on the whole much disappointed but things may look up yet.

22 & 23 6/24

Peaceful days in camp or a mile or two away sketching [*Pls 83–5*] & writing letters – rained each p.m.

24 6/24   *Hattu*

Geoff, Shebbeare, Hingston, Odell & Beetham started on 4 days' outing – down the gorge to Nepal border.

    Somervell & I with 3 porters climbed some 2000 feet above Tropdé to a most delightful spot – a flat shelf on hill – with a village (~~Hattu~~ Tingsang) & some cultivation. Here we camped. Got glimpses of top of Gaurisankar, a lovely summit high in mid air. Heavy shower in night.

Pl. 83 (L)   Rongshar Valley. Chuphar. 23 6/24.

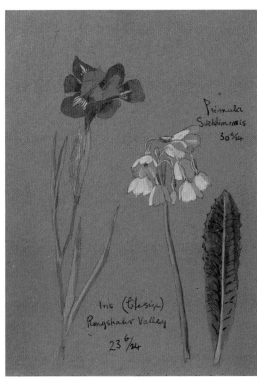

Pl. 84 (S) Iris (lifesize). Rongshar Valley. 23 6/24. Primula sikkhimensis. 30 6/24.

Pl. 85 (S) Rongshar Roses. Our camp covered with both sorts + the iris on preceding page carpeted the ground. *[probably 23 6/24.]*

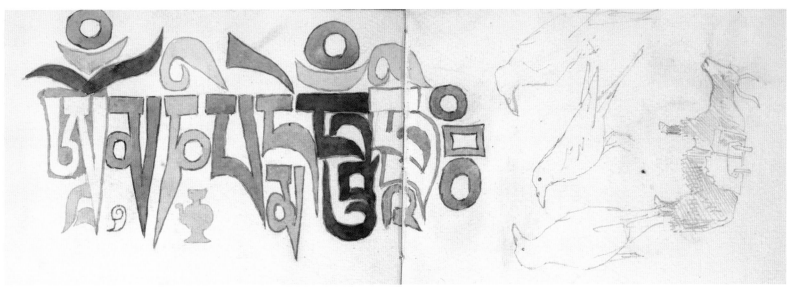

Pl. 86 (S)   *Om mani padme hum* [*Hail, jewel of the lotus flower*].

25 6/24   *Hattu*

Gorgeous views up to 8 a.m. – the most splendid snow peak I ever saw. Later climbed to pass 3000 feet higher & got some even grander views tho' only glimpses. Both sketched hard [*Pl. 89*]. Flora & fauna new & attractive – we ought to have made main camp here. Heavy shower in night again.

Pl. 87 (S)   Trintang. *Probably 24–25 6/24.*

LEFT: Pl. 88 (L) First view of Gaurisankar from Tingsang. 24 6/24.

RIGHT: Pl. 89 (L) Gaurisankar. 25 6/24.

Pls 90–1 (L)   Gaurisankar. 26 6/24. Night's candles are burnt out and forward day Stands tiptoe on the misty mountain tops. *The quotation is from Romeo and Juliet.*

26 6/24	*Tropdé*

Returned along hill & descended straight in camp. Wonderful views early – bathed & read English mail.

27 6/24	*Tropdé*

Quiet day – walked beyond Chuphar in p.m. – gorge party back to lunch.

28 6/24	*Tropdé*

In a.m. walked some miles down gorge with Shebbeare.

29 6/24	*Tropdé*

Wet almost all day – wrote letters & Times communiqué.

30 6/24	*Tang Chu*

Walked all the way – uneventful day.

1 7/24	*Tazang*

Ditto.

Pl. 92 (S) Burhel at Tazang. 1 7/24.

Pl. 93 (S) Geoff Bruce. Shebbeare. Transport. Rongshar Valley. 1 7/24.

OUR PICTURE PUZZLE PAGE
Who is this?
What is he reading?
How long has he been here?

Pl. 94 (S) OUR PICTURE PUZZLE PAGE. Who is this? What is he reading? How long has he been here?

Pl. 95 (S) Pusi La + Odell reading the [*Times*] Literary Supplement [*Pusi La crossed on 19 6/24 and again on 3 7/24*].

Pusi La

+Odell reading the Literary Supplement

Pl. 96 (S) Tingri Plain from Sharto. 4 7/24 [*Dated 4 4/24 in error*].

2 7/24	*½ way to Kyabrak*

Ditto – delightful camp, keen air – rain.

3 7/24	*Kyabrak*

Walked all the way crossing Pusi La (17500) with ease – feet doing pretty well.

4 7/24	*Shartö*

Long march (rode) – lovely views of Tingri plain towards end.

Tingri Plain from Sharto 4 7/24

Pl. 97 (S)   Kyang and Gazelle on Tingri Plains. 5 7/24.

Pl. 98 (S)   Shebbeare. 'It was grand to see that mountain horseman ride.' *The quotation is from 'The Man from Snowy River' by Andrew Barton 'Banjo' Peterson.*

Pl. 99 (S)   *Sketches of local people.*

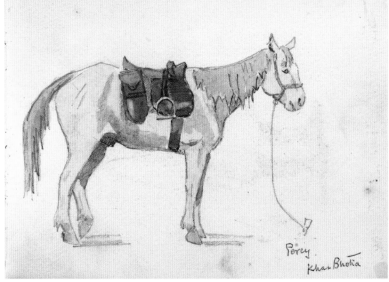

Pl. 100 (S)   Percy. Khan Bhotia.

5 7/24   *Tingri*

Short march (rode) – pleasant camp on river bank – rain. Saw gazelle, kyang, Tibetan sand grouse. Again wonderful views & colouring – rain all round & considerable wind in camp.

6 7/24   *Memo*

Pleasant march – pleasant camp – rain.

7 7/24   *Shekar*

Long march via bridge – got all business with Dzongpen settled before transport arrived. Must wait an extra day for transport.

8 7/24

Fairly busy day in Shekar – sorting stores, entertaining J.P. &c. Annoyed to find Hazard has gone to Tsang Po contrary to J.P.s wishes.

Pl. 101 (S) Southwards from Shekar Dzong. 8 7/24.

9 7/24   *Kyeshong*

Late start after heavy & prolonged rain all night up to 8 a.m. Transport
cut down to 100 animals thanks to Geoff.

10 7/24   *Trangsoe Chumbab*

Walked about ½ way with T.H.S. Then rode – rain after we got into
camp.

11 7/24   *Jikyop*

Rained all night until 8 a.m. Late start – Phung Chu enormous –
transport not in till 5 or 6 p.m. Heavy rain as they got into camp. What
a monsoon! Walked ½ march – feet very nearly O.K.

*Pl. 104 (L)   A page of finished pencil drawings, several of them based on
earlier sketches, including head of Shebbeare.*

*Pl. 105 (L)
A worked-up
painting of an
old man in local
costume based on
sketches in Pl. 99.*

Pl. 102 (S)   The Pirate King.

Pl. 103 (S)   An old man in local costume.

12 7/24   *To*

Long job crossing Puchung Chu – had to manhandle all loads. Transport drivers & our men did very well. Only settled into camp at dusk.

13 7/24   *Gangra*

Walked over Bahman Dopté – from stream to pass 1 hr. 20 min. without an effort. Then rode to Tso mo tre tang lake for a late tiffin – weather clearing. Lovely views. Into camp at dusk. 30 mile day.

14 7/24   *Tinki*

Walked all the way – v. short march – Tinki valley looking lovely. Interviewed Jong pens – strafed them about dak – no change – transport no trouble. Lovely day.

Pl. 107 (L)   Tso-Mo-Tray-Tang. 14 7/24.

Pl. 106 (S)   Sankar Ri from To. 12 7/24.

Pl. 108 (S)   Kamba Dzong. 18 7/24 [*in error for 16 7/24*].

15 7/24   *Linga*

Walked all way – L. foot sore. Jolly march – some rain.

16 7/24   *Kampa Dzong*

Met Hazard.

17 7/24   *Tatzang*

Hazard, Shebbeare & Odell went off to Gantok. We got caught in heavy hailstorm on Sung Dinghi La.

18 7/24   *Between Chaghun & Donka La*

V. cold & lots of rain.

19 7/24   *Near Tang La*

V. wet p.m.

20 7/24   *Phari*

Marched in torrents of rain all day. Got everything satisfactorily fixed for next day's march.

21 7/24   *Gautza*

V. pleasant march – rain latter half. My L. foot d-d sore again.

Pl. 109 (S) Om mani padme hum [*Hail, jewel of the lotus flower*]. Mani wall near Tatsang. 17 7/24.

Pl. 110 (S) Singdingi La. [*17 7/24.*]

Pl. 111 (S) Meconopsis or Blue Poppy. Chagun La. 18 7/24.

22 7/24   *Yatung*

Hospitable reception by Noels & Macdonalds. Work in p.m.

23 7/24   *Yatung*

Squaring up, shoeing horses &c.

24 7/24   *Yatung*

Mostly social – lunch party in big tent. My foot bad – blood poisoning.
Sold all stores satisfactorily.

25 7/24   *Chumbi tang*

Rode all way, foot very swollen, v. jolly march & nice bungaloo.

26 7/24   *Karponang*

Rode all way, over Nathu La. Perfectly gorgeous flowers & a most
delightful march up to pass. 8 or 10 kinds of primula alone. Rain
began on pass & continued all day. Passed Tsamgu – beautiful lake but
shrouded in mist. More delightful alpine country just below it. Finally
reached deciduous sub-tropical jungle. Long march say 21 miles.

27 7/24   *Gangtok*

Rode all way, foot still bad. Sunny morning – lovely views down valley
– true Sikkim blue.
   At Gangtok called Maharajah & Buckner his P.A.

Pl. 112 (L)   Phari Dzong. 21 7/24.

Pl. 113 (L)   Phari [*probably 21 7/24*].

Pl. 114 (S)   Goodbye to the Plateau. 21 7/24. Primula sikkimensis + Pink Polygonum.

Pl. 115 (S)   Siniolchum from Gangtok. 28 7/24.

FAR LEFT: Pl. 116 (S)   Geoff washing his feet at Sija [or *Siju*].

LEFT: Pl. 117 (L)   Nearing civilisation [*Geoff Bruce*].

Pl. 118 (L)   *View of the foothills, perhaps looking towards India.*

28 7/24   *Pakhyong*

Typical Sikkim – just like Sedongchen Rangli march. Rode all way. Heavy rain. 11 miles.

Dispatched all heavy kit in carts to Kalimpong road.

29 7/24   *Pedong*

Uneventful & not v. interesting march. I miss the high country. Rode all way, foot still pretty bad.

30 7/24   *Kalimpong*

Called on Perrys & Dr Graham called on us. All remaining heavy kit to Kalimpong road in cart.

31 7/24   *Lopchu*

Started early, breakfasted with Listers.

1 8/24   *Darjeeling*

General B. met us with motors at 6th mile – so back to Darjeeling.

THREE LETTERS FROM THE 1924 EXPEDITION

Two letters of condolence

Back at Base Camp, Norton took time out from organising the expedition's departure and composing communiqués and official reports to write letters of condolence to Mallory's widow, Ruth, and Irvine's father. The first, largely factual paragraphs of the two letters are worded in similar terms, but contain some differences in his thoughts on Mallory and Irvine's death. There follow some poignant personal appreciations of the two men, which complement the public tributes printed later.

Norton's comments on the fate of Mallory and Irvine carry the authority of one who had climbed to nearly 27,000ft with Mallory and Somervell in 1922, who had himself been saved by Mallory from a potentially fatal slip on the descent that year, and who had reached to within a thousand feet of the summit just a few days before Mallory and Irvine were lost. He knew the upper reaches of the mountain as well as anyone else alive, and he knew the conditions near the summit that year. Already at Camp III on 9 June, whilst anxiously awaiting news from higher up the mountain, he had confided to his Diary: '10 to 1 they have "fallen off" high up.' This remained his conviction, for reasons which are briefly set out in the letters of condolence. It was the view of the majority of the expedition, whom he assembled to discuss the issue before they left Base Camp. Only Odell believed that they had been benighted on the mountain, on their way down from the summit. Norton's view, and his suggestion that the accident was more likely to have occurred on the way down, was dramatically confirmed when Mallory's body was discovered in 1999.

As to the likelihood of Mallory and Irvine having reached the summit, in the letters of condolence he puts it an even chance. In the expedition book he concluded that it was 'not proven', and 'that is all that can be said about it'. The discovery of Mallory's body has not resolved the question; nor has it dampened speculation. In view of the suggestion that has sometimes been made that Mallory would have been unable to resist pressing on to the summit, even in the face of insurmountable odds, it is worth reading Norton's words to Ruth Mallory written just a few days after the event.

> *Mt. Everest Expedition*
> *Rongbuk Base Camp*
> *13-6-24.*

Dear Mrs Mallory,

The news of your husband's death will have reached you long before you get this – firstly, if my arrangements don't miscarry, through the R.G.S. [Royal Geographical Society] – then with the brief outline of the facts in my communiqué to the Times of 11th June, & finally I think you will have seen the full story in the next communiqué to be sent off about 5 days hence. I cannot anticipate this method of giving you the full facts by any private letter as the Times communiqués are wired in press code from Simla and cross Tibet by special relay post.

As to the facts and circumstances I can add very little without repeating what you will have got from the above sources.

Everything points to the probability of a sudden death – a slip by one or other – a purely mountaineering accident. It is hard to invent any hypothesis which will cover the facts entailing the idea of a lingering death from exposure, nor is there any reason to suppose that any defect in the oxygen apparatus could have been the cause.

Whether or not the party reached the top must always remain a mystery: I put it at an even money chance. They were unaccountably late at the point where they were last seen and yet had time to get to the top without serious risk. They had apparently surmounted the most serious obstacle: on that particular ground the chances of a fall were greater on the way down than when ascending. There was sufficient mist and cloud on the day in question to account for our never seeing them again (despite the close watch by telescope that we kept on the mountain) even if they had made the summit and turned back again.

There must be many points you want to know. I wish I could foresee and answer them now; I shall of course come and see you – if you wish – as soon as I get home – probably in October.

And now I wish I could help you in your great grief – or even express adequately my own sorrow and sympathy.

You know that I was always a friend of that gallant gentle soul – but you can hardly know how close this last expedition has drawn us together.

Owing to his joining the expedition so late I always felt that I had usurped the place he would have otherwise have filled – I was always conscious of the inferiority of my qualifications to do so and guessed that he might feel the same. Yet never once by word or deed or hint did he suggest such a thing. He simply played up to me and backed me up through thick and thin.

In every detail of the mountain campaign we worked in together and his unfailing sound advice, his organizing ability, his bottomless capacity for work and his determination to win were my prop and stay.

You can't share a 16 lb. high altitude tent for days and weeks with a man under conditions of some hardship without getting to know his innermost soul and I think I know almost as you do what his was made of – pure gold.

And then – on the mountain – I wish I could describe to you what he was like. Physically he was a wonder – the best of us without the least doubt from start to finish – but this hardly mattered for his great heart carried him on entirely independent of physical considerations. I really believe the struggle between him and the mountain had become a personal matter to him. He simply would not accept defeat and yet (from 1000 talks on the matter) I know how his determination was tempered with discretion; he fully realised his responsibility as leader of the climbing party – he and I saw eye to eye over the question of the absolute necessity of avoiding a single casualty even to conquer the mountain – and he has often told me his views as to the point at which the leader of a party must turn back for safety's sake however near the goal.

He was a great mountaineer.

I have written so far rather from a personal point of view, but I can assure you my sentiments are shared by the whole expedition.

He compelled the admiration of all of us – but he was also a real "pal" to all. I won't say I never heard him say a rude word to anyone – he was too much of a man – and he cursed Somervell and me roundly for lagging or loitering when there was serious business in hand on at least two occasions – but it was worth it for the warm hearted apology which always followed so quickly.

I am keeping any of his private kit which can be of any use to you and shall send it you in due course.

What a tragedy it is! If only I could help you in some way.

Yours sincerely

E.F. Norton

P.S. My best address is c/o Cox & Co Bombay until Sept. 7th and then Uplands, Fareham, Hants. This is the nearest I can say. But please don't answer this unless you have something to ask.

✳ ✳ ✳

Mount Everest Expedition
Rongbuk Base Camp.

13-6-24.

Dear Mr. Irvine,

Unless the steps I have taken have miscarried you will long before this have received the news of your son's death – I hope first from the Mt. Everest Committee and then with full details in two successive communiqués to the Times.

I could not anticipate the latter method of communicating the news in full by a letter as the Times have a special relay post across Tibet and then wire their communiqués home by press code.

As to the facts and circumstances I fear I can add but little to what you will have heard from the above sources and a letter you will get from Odell.

Everything points to the probability of a sudden death – a slip by one or other – a purely mountaineering accident. Personally I cannot suggest any hypothesis to cover the idea of a lingering death from exposure, nor is there any reason to suppose that the cause might have been due to a defect in the oxygen apparatus for Odell's experience proves that for people as fully acclimatised to altitude as was your son oxygen may be entirely dispensed with as soon as the descent is begun.

Whether the party reached the top or not must always remain a mystery. I put it myself at a very even chance. They were unaccountably late at the point where they were last seen by Odell – but not too late to reach the top in time to return safely. They were reported as "going strong." On the particular ground in question a slip was more likely to occur descending than ascending for they had apparently surmounted the most serious obstacle in the ascent. The pair, of course, hold the world's altitude record.

There must be so many points that you want to know. I shall be only too delighted to answer any by letter or to come and see you on my return to England – probably in October. The nearest I can give you as an address is c/o Cox & Co. Bombay until say Sept 7th and then Uplands, Fareham, Hants.

I wish I could in any way help you in your great grief or adequately express my sympathy.

In the sort of experience I have shared with your son one gets to know people better in 6 weeks than often in 6 years of easy home life, so both from my own knowledge of him and from much that Odell has told me I can guess what your son was to you.

To me the whole thing is very bitter, my fixed determination was to bring off success if possible, but, success or failure, above all things to avoid casualties – and I thought it could be done. I was determined that such splendid lives as those we have lost were infinitely too high a price to pay for success.

Much that your son was to us I have already written of in various communiqués to the Times. From the word "go" he was a complete and absolute success in every way. He was spoken of by General Bruce in an early communiqué as our "experiment". I can assure you that his experimental stage was a short one as he almost at once became almost indispensable. It was not only that we leant on him for every

conceivable mechanical requirement – it was more that we found we could trust his capacity, ingenuity and astonishingly ready good nature to be equal to any call. One of the wonderful things about him was how, though nearly 20 years younger than some of us, he took his place automatically without a hint of the gaucherie of youth, from the very start, as one of the most popular members of our mess.

The really trying time that we had throughout May at Camp III and the week that he put in at Camp IV in June (of which I spoke in the last communiqué before I knew of his death) were the real test of his true metal – for such times inevitably betray a man's weak points – and he proved conclusively and at once that he was good all through. I can hardly bear to think of him now as I last saw him (I was snowblind the following morning and never really saw him again) on the N.Col – looking after us on our return from our climb – cooking for us, waiting on us, washing up the dishes, undoing our boots, paddling about in the snow, panting for breath (like the rest of us), and this at the end of a week of such work all performed with the most perfect good nature and cheerfulness.

Physically of course he was splendid – as strong as a horse. I saw him two or three times carry for some faltering porter heavier loads than any European has ever carried here before.

He did the quickest time ever done between some of the stages up the glacier. One of his feats was to haul, with Somervell, a dozen or so porters' loads up 150 feet of ice cliff on the way to the N.Col.

As for his capacity as a mountaineer the fact that he was selected by Mallory to accompany him in the last and final attempt on the mountain speaks for itself.

I hope you will express my deep sympathy and regret to Mrs. Irvine, and to your sons and daughter. Please write to me if I can give you any information or help you in any way – otherwise do not trouble to answer this letter.

Yours sincerely,

E.F. Norton.

Letter to Sir Francis Younghusband

During the expedition, Norton corresponded with the Mount Everest Committee in London through the Committee's secretary, Arthur Hinks. He also received regular personal letters from Sir Francis Younghusband, the Committee Chairman, in which Younghusband supported Norton through the repeated setbacks on the mountain and complimented him on his leadership in the face of adversity. In this letter, written the day after the return to Darjeeling, Norton thanks Younghusband for his support, reflects on the lessons to be drawn from the 1924 expedition and looks forward to another attempt in the future. He ends with an amusing account of the expedition's reception at Darjeeling and the contrast they presented to the pallid faces of the local expatriate community.

✻ ✻ ✻

Hotel Mount Everest.
Darjeeling Aug 2nd 1924.

Dear Sir Francis.

I think I have two letters from you to acknowledge – of July 3rd & 8th respectively.

I am delighted to find from the article you wrote to the Times that you are so entirely in agreement with the views of myself & all of us – that now is not the time to talk of giving up the attempt to climb the mountain. What you say as to the stiffening of public opinion – or rather the strengthening of that section of it which has always regarded the enterprise favourably – exactly confirms what I anticipated: it is so typical of our attitude as a people.

In fact I have tried to say something of the sort in the last communiqué I wrote to the Times: I was not quite sure if the Committee would endorse the views I therein expressed & I am more than pleased to hear that you at least do so so heartily.

What you suggest about epitomising what has been done this year will be done by the General as he is taking on the winding up dispatch from here. I think a little fresh blood – or ink – will be a good thing for

the demand for 15 communiqués coming so late has about exhausted my particular fountain pen & I'm afraid the series of communiqués after the climax was over may savour of flogging a dead horse.

Your letter of 8th was – as every one of your letters has been – very cheering to me – though I confess I still feel that you are all inclined to overestimate our achievements at home. For I feel that it was a foregone conclusion that – provided we could sleep twice above 23000 – the 1922 record must be handsomely beaten.

That we did so sleep is the achievement of our fine porters (I confess that the one thing I privately pat myself on the back for is that I had the influence over our little lot – & just sufficient Khaskura – to bring them to the sticking point the morning we woke at 25000).

It's a curious psychological fact that here are the men – on the spot – who could climb Everest without turning a hair.

Yet the sahib – so physically inferior in this respect – has to come 1000s of miles to give them a lead – even to 27000.

What you say about the standard of achievement is a great theme of mine.

My grandfather never climbed Mont Blanc without the most exaggerated symptoms of mountain sickness – the like of which I have never seen on Everest: while my sister & I hardly felt the altitude on the final ridge of Mont Blanc last year. In fact my grandfather was often mountain sick at little over 10000 – or thought he was.

And already Odell & Hazard have been for a 'constitutional' to the height (25000) that three of us reconnoitred in '22 to ascertain if men could live there. The N. Col is now regarded as the least to be demanded of every porter or climber in fine weather. I slept really well at 27000 & one will expect everyone else to do so in future.

I find one misconception current here due partly to a misprint in Somervell's dispatch, everyone thinks that about 28000 is the limit of human endurance. The papers say I took an hour to struggle 8 feet above Somervell – a truly moving picture of heroic determination.

The 8 should of course be 80, besides say 500 yards horizontally which S. didn't mention, but these figures mean nothing, for I left Somervell on the last of the really easy going & almost at once got onto very narrow sloping slabs which, especially when snow covered, needed the greatest caution & I once or twice had to retrace my steps: the slowness of my progress, compared even to what we had done up to then, was entirely attributable to this: on similar ground to what we had traversed before I should have covered a good deal more than twice the height or distance.

I mention this because on reading the dispatch again I quite see that there is ground for the idea that one more or less ran down like a clock.

Somervell himself gave out simply on account of his cough. Later the same evening I believe he nearly pegged out – for he was actually half choking when he coughed up a lot of blood & stuff out of his throat – & this relieved & enabled him to get his breath again.

Referring to your last letter again, I shall of course be only too delighted to come & spend a night with you at Westerham, for I should like to have a good yarn about it all.

I think I shall probably creep home via Havre & Southampton with my people from the Alps – for we live near Southampton.

This will probably be either at the end of September or, if I get a month's extension of leave that I have asked for, perhaps 10 days later. I shall then come up to town soon.

The General, Geoff & I are all coming home in the Narkhunda together so, if we get a line to go on, we can fix up what we are all going to say at the Albert Hall without necessarily meeting again.

I hope I shall be let down easy on Oct 17th as I am an entire novice at the game; I have only lectured to soldiers & sailors – simple folk with a not very exacting standard.

I should prefer to confine myself to introducing this year's party, as I want this very fine team to be flesh & blood to the British public as far as possible.

We all arrived here yesterday – heavy baggage via Kalimpong road & the railway, Hazard from Gangtok & the rest of us from Kalimpong & Lopchu: a final triumph of staff work by Geoff – synchronisation on the battle field.

Odell & Shebbeare get in tomorrow.

The porters were marched through the streets behind the police band & I think very much appreciated their welcome. We had hardly any but the best with us; the few 'duds' were conspicuous at the head of the procession, mostly on horseback: the real toughs who carried to 27000 mostly slunk inconspicuously in rear – such is human nature – Bhotia & otherwise!

The General met us with motors; he is looking A.1. & full of cheer. We have had tremendous yarns about everything of course.

Those of us who have arrived so far average some 8lbs under weight – I guess a stone when we left Rongbuk – but we are all bursting with fitness & the faces of the party were a pleasant contrast to the prevailing complexions at this tamasha yesterday, for they have hardly seen the sun in Darjeeling.

With very kind regards to Lady Younghusband.

Yours very sincerely,

E. F. Norton.

AFTER EVEREST

Many years later the distinguished Everest climber, Eric Shipton, summed up the 1924 expedition: 'the story of that expedition is one of extraordinary perseverance in the face of crushing misfortune, of outstanding achievement against heavy odds, of success so nearly grasped, and, finally, of tragedy. Throughout, the rare qualities of Norton's character, both as an individual and as a leader, are clearly revealed'. In spite of all that had been achieved, Norton returned to Darjeeling deeply disappointed at the expedition's failure to reach the summit, and at the loss of Mallory and Irvine, yet confident that the mountain could be climbed, and eager to have another chance at it. In the event, this was not to be: the Tibetan government refused permission for further expeditions.

Norton returned to Europe, and this time was caught up in the public aftermath of the expedition. On 17 October 1924 a memorial service for Mallory and Irvine was held at St Paul's Cathedral, followed later the same day by a joint meeting of the Royal Geographical Society and the Alpine Club at the Royal Albert Hall. At this, rather reluctantly in view of his inexperience as a public speaker, Norton spoke of the individual members of the expedition and described his and Somervell's record-breaking climb without oxygen. The proceedings were published, as was a lecture he gave to the Alpine Club on 15 December in which

he summarised the lessons which had been learnt from the 1922 and 1924 expeditions and assessed the prospects for the future. He also wrote large parts of the official expedition book, *The Fight for Everest, 1924*, which appeared under his name in 1925. With these duties performed, he resumed his military career. He married Joyce Pasteur in December 1925. When another attempt on the mountain was eventually launched, in 1933, Norton was approached to lead the expedition, but his professional obligations did not permit of a prolonged leave of absence. In spite of many years as a senior army officer in India, he never attempted any further high climbs in the Himalayas, preferring to spend as much time as he could in the mountains around his beloved family chalet above Sixt.

In 1925 Norton was elected an Honorary Member of the French Alpine Club for his exploits on Everest. The following year he was awarded the Founder's Gold Medal of the Royal Geographical Society for the very distinguished leadership shown during the 1924 expedition, as well as his personal achievement of ascending higher than had been climbed by any other man who had lived to tell the tale. The citation congratulated him 'upon the high spirit and courage with which he attacked the mountain, and also upon the self-effacing temper in which he conducted the expedition to the verge of absolute success'.

The public accolades can be fleshed out with the comments of those who worked with him during the two Everest expeditions. Odell recorded that 'Norton was a delightful person to travel with, not merely a first-class soldier ... and a very good painter'. Mallory, in a letter to the Mount Everest Committee in 1924, was explicit:

> I must tell you, what Norton can't say in a dispatch, that we have a splendid leader in him. He knows the whole bundobust from A to Z, and his eyes are everywhere, is personally acceptable to everyone and makes us all feel happy, is always full of interest, easy and yet dignified, or rather never losing dignity, and is a tremendous adventurer.

Somervell described the qualities of leadership shown by Norton in these words:

> When he took over the leadership in 1924, he showed himself the ideal leader. He continually asked us all what our opinions were, and respected them, and then led in such a way that we all felt we had our hand in his, and were as it were yoked to him. Yet all the time he knew what was best for us and got us all to do it without any bossing or ordering – he pulling his side of the yoke while we pulled ours, and all in complete harmony.

Sir Francis Younghusband singled out the particular importance of the spirit of comradeship which had sprung up between the European climbers and the Himalayan porters. General Bruce, he said, with his deep knowledge of the Himalayan peoples, had led the way. But on Everest itself:

> it was Norton who set the standard and established the code ... Especially had he engrained in him the principle that a leader must look after the least of every one of his followers before himself ... And indeed he did stand by them on that notable occasion when four of them were marooned on the ice-face of the North Col, and he, Somervell and Mallory, at the risk of their lives, and at the risk of the success of the whole expedition, climbed up perilous

ice-slopes to their rescue. By these and other deeds Norton and the other climbers established the tradition that these Himalayan porters were something more than carriers: they were fellow-mountaineers, comrades in high enterprise.

Younghusband particularly praised Norton's success in urging three of the porters to climb for the first time beyond Camp V and help establish Camp VI. Without this, neither his own attempt on the summit with Somervell, nor Mallory and Irvine's final climb, would have been possible. This achievement appears all the more remarkable in view of what emerged during a chance meeting in Darjeeling in 2002. Norton's son Bill, accompanied by the 1953 Everest veteran, George Band, met Ang Tsering, a 97-year-old former Everest porter who had been on the 1924 expedition and had actually been one of the 'Tigers' who helped Mallory and Geoff Bruce establish Camp V at about 25,000ft, but would go no further. He revealed previously unknown details of the expedition's visit to the Rongbuk Monastery following the crisis on the East Rongbuk glacier. The Head Lama, he recalled, had told the Sherpas in their own tongue to do whatever the sahibs asked them, 'but not so that they reach the summit, because this would be a bad omen for your families, for the monks here at Rongbuk and for the local people'. The expedition had been wrestling, unknown, with forces beyond those of the elements!

In his posthumous appreciation published in 1955, Eric Shipton reflected on Norton's achievements on Everest from the perspective of a further three decades' experience of the formidable challenges presented by the mountain. Discussing the 1924 expedition, he wrote:

> Climbing to more than 28,000 feet without the help of oxygen apparatus was a very fine achievement. For men in the condition that Norton and Somervell were in, after weeks of unremitting hardship and frustrated toil at altitudes which sap the strength of the strongest like wasting disease, their feat was a prodigious display of courage and determination. Many of us believe that if

they had been spared those earlier fruitless efforts, they would have reached the top. Norton's record on Everest won him lasting fame in the annals of human endeavour.

No-one who met General Norton could fail to be impressed by his personality, his quiet authority, his warm sympathy, and above all his integrity. He inspired immediate respect and affection. I count it one of the chief rewards of my association with the Everest adventure that through it I was privileged to know him.

Although in the end the 1924 expedition to Everest had ended, in Younghusband's word, in a repulse, it had demonstrated beyond reasonable doubt that a way to the summit could one day be found. Norton himself remained convinced that the summit could be won. Reviewing the situation in an article published in May 1950, four years before his death, he suggested that success was most likely to be achieved by placing a *third* high camp as close to 28,000ft as possible. He also continued to believe that the summit could be reached without oxygen by men who were sufficiently physiologically gifted for high altitude climbing.

The latter prediction was not fulfilled until Reinhold Messner and Peter Habeler reached the summit without supplementary oxygen in 1978. But it was only three years after his article that Edmund Hillary and Tenzing Norgay, using oxygen, finally reached the summit of Everest after spending the final night in a camp pitched just short of 28,000ft. The experience gained by the early expeditions had not been for nothing.

The final word may be given to the leader of the successful 1953 expedition, John Hunt. Shortly after his return to Kathmandu from Everest in June 1953, he wrote to Norton. Stressing the importance of Norton's advice on the matter of the final camp before the summit, he added:

I would also like you to know how very much you and the 1924 Expedition have been in our minds during recent months. We have always regarded your team as a model which we should emulate; as you know, we have throughout been very conscious of the huge debt which we owed to you and to others who have gone before us.

To you goes a big share in the glory.

NOTES

For ease of reference, the official expedition books are referred to in the Notes as *Everest 1921*, *Everest 1922* and *Everest 1924*. Full details of these and other publications can be found under Further Reading.

THE 1922 EXPEDITION

The 1922 Diary

27 3/22   The main body of the expedition, led by General Bruce, had left Darjeeling on the 26th. Travelling by train south to Siliguri and then back up the Teesta Valley branch line, they took two days to reach Kalimpong. Norton and his companions, taking the road east out of Darjeeling, reached Kalimpong in half the time.

28 3/22 and ff.   From Darjeeling to Shekar Dzong the expedition followed mostly in the footsteps of the 1921 reconnaissance expedition, and the route is described more fully in *Everest 1921*. They travelled from Kalimpong in two parties, the first of which (including Norton) set off on 28 March.

3 4/22   Longstaff gives a memorable account of the hunt for the ibis-billed curlew in the expedition book (*Everest 1922*, 326–7),

culminating in the two men wading waist-deep across the icy stream to recover the specimen.

4 4/22   'Bird notes': Norton kept separate notebooks with his observations on birds.

The second party reached Yatung on the evening of the 3rd, but the first party went on ahead again on the 4th. The two parties joined up at Phari on the 6th.

The Bérard valley is in the French Alps near the family chalet above Sixt.

9 4/22   The 1921 expedition had continued northwards from Phari to the Dug La and then westwards to Kampa Dzong, whereas the 1922 expedition took the more difficult but shorter direct route north-west out of Phari.

9–10 4/22   Angzang Trag is called Hung-Zung-trak in the expedition book (*Everest 1922*, 36).

5 and 6 4/22   In both entries a space was left to add the name of another of the mountains, but was not filled in.

11 4/22   Dr Alexander Kellas was a pioneering Himalayan explorer and climber who reached a height of 23,400ft on Mount Kamet in 1920. He joined the 1921 reconnaissance expedition, but was taken ill and died on the journey out. He was buried at Kampa Dzong.

Captain Farrar, along with Mr Meade, had overseen the provision of equipment in London. He ordered items individually tailored to the needs of the different members of the expedition, including, evidently, an extra-large sleeping bag for Norton.

13 4/22 Finch and Crawford had stayed behind at Darjeeling, awaiting the arrival of the oxygen equipment, which had been delayed. They left Darjeeling on 2 April.

16 4/22 Norton may have learnt the value of a catapult for obtaining specimens from young Macdonald: a letter to London written on the return journey refers to the contribution to the natural history collections made by Macdonald's 'unerring catapult'.

20 4/22 The village of Mendé lies at the crossing of the Ko Chu or Yaru river near Linga (*Everest 1921*, 57). The expedition had crossed there on 15 April – hence the reference here to the *previous* crossing of the river.

The name of the mountain, Sangkar Ri, was added later into a space left in the original entry.

20–1 4/22 The hitch with the transport at Gyanka Nangpa is mentioned again in the diary entry for 18 4/24.

24 4/22 Upstream of its junction with the Yaru near Shiling, the Arun is generally known as the Phung Chu.

27 4/22 On leaving Shekar Dzong, the expedition diverged from the route taken in 1921. The previous year's reconnaissance expedition had continued westward to Tingri Dzong and commenced exploring the approaches to Everest from the west. Thanks to the geographical knowledge attained in 1921, the 1922 expedition was able to approach Everest by a much more direct route from the north, some of it through previously unexplored country.

30 4/22 The Guivra stream is an Alpine stream not far from the family chalet above Sixt in the French Alps.

1 5/22 Tuppoo is presumably the Tibetan private servant engaged by Norton who is described as a very capable cook in the expedition book (*Everest 1922*, 24).

1 5/22 and subsequent entries The 17000ft Camp is Base Camp.

2 5/22 The 'home mail' refers to the letter to his mother of this date printed on pp. 57–8.

5–9 5/22 These entries describing the exploration of the East Rongbuk glacier and the establishment of Camps II and III were written up retrospectively back at Base Camp on 9 May. It is probably for this reason that some confusion has arisen as to which night was spent at which camp. Careful reading of the diary shows that the nights of 5 and 6 May were actually spent at Camp I; the nights of 7 and 8 May were spent at Camp II; while the night of 9 May was spent back at Base Camp. This is confirmed by the second of Norton's letters home (see pp. 59–62), which describes the whole episode at greater length.

5 5/22 The site of Camp I, including a couple of sangars, was estimated in 2001 to lie at a height of about 18,100ft (Hemmleb and Simonson, *Detectives on Everest*, 73).

6 5/22 Wheeler's 19,360ft photographic station. During the 1921 reconnaissance, extensive survey work had been carried out by Morshead and by Oliver Wheeler, a Canadian officer. He employed a special photographic survey technique, and in 1921 he had surveyed part of the East Rongbuk glacier, though no-one before 1922 had explored its entire length. His 19,360ft photographic station is shown on the central sheet of the map of Everest published in 1922 on the basis of the previous year's reconnaissance, on the west side of the East Rongbuk glacier (see Fig. III). See also below, Note to Letter of 10 5/22.

Poo was one of the porters. He participated in all three of the early Everest expeditions.

8 5/22 Oxygen bottles and other remains of the 1922 Camp III were found on the upper East Rongbuk glacier in 2001 at a height of about 21,000ft (Hemmleb and Simonson, *Detectives on Everest*, 65 and 79–81).

9 5/22 Uncle Jack was Jack Wills, son of Sir Alfred Wills, the pioneering Alpinist and builder of the family chalet above Sixt. Jack was one of the older generation who taught Norton to climb during the annual family visits to the chalet. A member of the Alpine Club, he may have played some part in Norton's selection for the 1922 expedition.

Norton's father, also called Edward, was an entrepreneur with business interests around the world. He died in May 1923.

10 5/22   The letter to his mother is printed on pp. 59–62.

14–18 5/22   These entries appear to have been written up daily on the East Rongbuk glacier at Camp II or III. The entries for 15–26 May use a different dating formula; the standard formula resumes on 27 May.

19–26 5/22   The account of the high climb appears to have been written up on 26 May, after the return to Base Camp. The description of the summit attempt in his letter home dated 25 May (see pp. 62–6) was therefore written before the corresponding diary entries, which in places echo the more expansive style of the letter.

20 5/22   In *Everest 1922*, 187 Mallory indicates that he was responsible for the loss of Norton's rucksack. Neither in the diary nor in the letter home does Norton attribute any blame to him. That night the two of them shared a tent at Camp V in great discomfort, described at length by Mallory in *Everest 1922*, 194–201. A more matter-of-fact but less well-known account of the summit attempt written by Mallory was published in *The Alpine Journal*, 34 no. 225 (November 1922), 425–39, reprinted in Gillman, *Climbing Everest*, 205–21.

21 5/22   The height at which the summit party turned back was initially calculated as about 26,750ft, using the aneroid barometer. Subsequent measurements using a theodolite calculated the altitude as 26,985ft (*Everest 1922*, 210).

'I then slipped badly …' This incident, which would have changed the course of mountaineering history had it ended differently, is described by Mallory in Gillman, *Climbing Everest*, 214 and in *Everest 1922*, 214–15. Mallory neither names the climber who slipped, nor mentions that it was he himself who saved the party. Somervell, in his account published some years later (*After Everest*, 66–7), notes that he himself had been at the rear while it was Mallory, leading down, who saved the party; but he too, in the manner of the time, did not see fit to name the initial faller. Some modern accounts have stated that it was Morshead who fell. Norton's testimony makes it possible to correct the record.

In 2014 an ice-axe said to be the one which Mallory used to arrest the fall was offered at auction (Christie's, London, sale of *Travel,* *Science and Natural History*, Thursday 10 April 2014, Lot 216). However, comparison with the famous photo of Mallory and Norton on their summit attempt on the same day (*Everest 1922*, pl. opp. p. 204) demonstrates that it is not the same ice-axe.

26 5/22   *Everest 1922*, 63 states, presumably in error, that John Macdonald arrived with a much-needed supply of money on 27 May.

27 5/22–4 6/22   A week in Base Camp covered in two brief entries, followed by two days with no mention at all, are probably symptomatic of Norton's exhaustion following his summit attempt.

5 6/22   On 3 June, Mallory, Somervell and Finch had left Base Camp for a final summit attempt. Finch, however, was still exhausted after his high climb with Geoff Bruce and was forced to abandon the attempt, having reached no further than Camp I.

11 6/22   The third attempt on the summit was to have been made by Mallory and Somervell using oxygen. This necessitated resupplying Camp IV on the North Col with heavy oxygen equipment and other supplies. On 7 June, Mallory, Somervell and Crawford were leading a large party of porters up the steep ice cliff above Camp III, which had a deep covering of fresh snow, when disaster struck.

13 6/22   'Old Father William' was the nickname given by Norton and Geoff Bruce to a very helpful, if rather officious old man who supplied the party with much-needed fresh vegetables (*Everest 1922*, 84 and 104).

16–17 6/22   After the disaster beneath the North Col, the remnant of the expedition, led by General Bruce, left Base Camp on 14 June and headed for the Kharta valley.

19 6/22   The picnic which Norton laid on is described by General Bruce in the expedition book as consisting of Gruyère cheese, sardines, truffled yaks and three bottles of champagne (*Everest 1922*, 88) – to which Norton, in his personal copy of the book, added the word 'quails'.

22–30 6/22   During this period Norton spent much of his time with Mallory, for the first time apart from when they had been climbing together. Norton was still nursing his frostbitten feet, while Mallory was stricken with grief and guilt over the deaths of the porters. The diary does not record what passed between the two, but this period of recuperation together may have laid the foundations for their mutual respect and strong working relationship, which were central to what was achieved in 1924.

3 7/22   Mallory, Somervell and Crawford headed directly from the Kharta valley to Tinki, and from there back to Darjeeling via Gantok in Sikkim. The rump of the expedition prepared to head north back to Shekar Dzong and retrace their steps with the remainder of the baggage train. The initial route from Lungmé to Shekar Dzong took them through previously unexplored territory.

11–14 7/22   A more northerly return route to Tinki was chosen so as to avoid having to wade through the Yaru twice and cross the high Tinki pass. 'To' is spelt 'Chiu' in *Everest 1922*, 107–8.

15 7/22   Even at the end of an arduous expedition to Everest, Norton was keen to spend time in the mountains at the family chalet near Sixt.

THREE LETTERS FROM THE 1922 EXPEDITION

Letter of 2 5/22

The Giffre is a stream near the family chalet above Sixt, which is dominated by the great mass of Mont Buet.

Letter of 10 5/22

'Wheeler's photographic survey point at 19,000ft.' In 1921 Wheeler explored about halfway up the East Rongbuk glacier from the main Rongbuk glacier. He was then forced to retreat, and travelled round to join the other members of the expedition on the Kharta glacier to the east of Everest. He crossed the Lhakpa La at the top of the Kharta glacier to reach the head of the East Rongbuk glacier beneath the North Col. The middle section of the East Rongbuk glacier was thus virgin ground. See also above, Note to diary entry for 6 5/22.

Camp II is said to be surrounded by moraine heaps or ice walls '700 feet high', according to the typescript. A marginal pencil correction '100?' is more plausible.

References to the Alpine landscape around the family chalet above Sixt include the Chardonnet, the glacier des Dardes, the Salle valley, the glacier d'Argentière and the Guivra stream.

Letter of 25 5/22

The height of 26,750ft was subsequently revised upwards to 26,985ft using a theodolite – see note to diary entry 21 5/22.

The Duke of Abruzzi had reached a height of about 24,600ft in the Karakoram near K2.

The col at the head of the East Rongbuk glacier overlooking the Kama valley is called the Rapiu La. See diary entry for 16 5/22.

Norton's opinion of the strawberry ice-cream which he improvised on the N. Col was not shared by Mallory – see *Everest 1922*, 221!

'Exhaustion in 1914.' Norton arrived in northern France in August 1914 and was appointed section commander of D Battery, Royal Horse Artillery in September. He was involved in intense fighting around the Franco-Belgian border in October and November 1914. For his part in these engagements he was mentioned in dispatches; soon after he was awarded the Military Cross.

THE 1924 EXPEDITION

The 1924 Diary

26 3/24   Norton arrived at Darjeeling early in March. On 13 March, General Bruce sent him with Shebbeare to Kalimpong to deal with the baggage that had been sent on there. Returning

to Darjeeling, he saw the main body of the expedition off on 25 March. He himself (as in 1922) set off a day later by motor and caught up with the rest of the party at Kalimpong in the afternoon. The Listers were tea-planters. 'D.B.' is short for Dak Bungalow, where they had spent the night in 1922.

27 3/24 'Same tamasha …' refers to another meeting with Dr Graham's scout troop. As in 1922, the expedition travelled from Kalimpong in two parties, the second being led by Norton. Lloyd, who set off with the first party for Pedong, was perhaps the unnamed young Welshman mentioned in *Everest 1924*, 20, who ran the scout troop. Mrs McDonald was the wife of David McDonald, the Trade Agent at Yatung whom Norton had met in 1922.

28 3/24 The Ranger is probably the young Sikkim military policeman mentioned in *Everest 1924*, 21.

2 4/24 At Chumbi the second party caught up with the first. John MacDonald, the son of the Trade Agent at Yatung, entertained the expedition in his father's absence (*Everest 1924*, 23). He again joined the expedition and accompanied them as far as Base Camp.

3 4/24 Sticking out the tongue was a local sign of greeting.

4 4/24 The Tibet authorities had been upset by the hunting and collecting activities of the 1922 expedition and had strictly forbidden any repetition in 1924.

4–5 4/24 The Dzong Pen of Phari attempted to charge inflated rates for transport, but eventually gave way.

6 4/24 One feature of the 1924 expedition was a vastly improved mess tent, described as 'Norton's special child' by Mallory, who waxed eloquent on its virtues (in spite of a lack of tablecloths) in two of his letters to his wife, Ruth (*Everest 1924*, 212–15). It is mentioned several times in subsequent entries.

7 4/24 Karma Paul was the expedition interpreter, Gyeljen his assistant. Kanza (or Kamcha) the cook died soon after the expedition's return to Darjeeling (*Everest 1924*, 82).

8 4/24 The 'Blizzard Camp' is referred to in the diary entry for 7 4/22.

10 4/24 Here and elsewhere Norton writes 'last year' when he means 'two years ago'.

14 4/24 '*Times* article.' The expedition was contracted to supply exclusive articles to *The Times* of London. These were one of the responsibilities of the expedition leader which Norton inherited from General Bruce. Bruce had sent the first two dispatches, so the first one sent by Norton was number three in the series. They are referred to regularly in the diary. The full list is given below under Further Reading.

16–21 4/24 It was decided to follow the northern route between Tinki and Trangso Chumbab which had been pioneered by the 1922 expedition on its homeward journey, i.e. following the Chiblung Chu.

17 4/24 The final agreement on a climbing plan for reaching the summit elicited an excited letter from Mallory to his wife, Ruth, describing his 'brain-wave' (*Everest 1924*, 219–20).

18 4/24 The incident with the transport at Gyanka Nampa is mentioned in the diary entries for 20–21 4/22.

19 4/24 Hingston accompanied the General most of the way back to Darjeeling and only caught up with the expedition at Base Camp on 11 5/24 (see diary entry for 12 5/24).

21 4/24 The expedition included four Gurkha non-commissioned officers, who had to be able to read and write simple messages in Hindustani written in Roman characters. One such message is copied into the diary entry for 2 5/24.

'After dinner …' The allocation of climbers to parties was one of the most delicate issues to be addressed, since it determined who was likely to have a chance at the summit, and which parties would use oxygen. The allocation was announced after dinner on 22 April (according to Mallory's letters) or 24 April (according to Norton's much later published account) (see *Everest 1924*, 45–6 and 221–3), but on 21 April according to the diary.

22 4/24 The 'cooly bundobust' refers to the logistics of transporting the necessary stores and equipment to the required locations on time, for which a detailed schedule of porter movements was worked out.

24 4/24 and 27 4/24 The thirty-six Tibetan coolies were an additional complement of porters engaged to help with setting up the lower camps.

Irvine recorded the gift of two oxygen cylinders to the head lama of the monastery at Shekar Dzong in his diary: 'we also told him there was a devil inside whose breath would kindle a spark – we showed him on incense' (Summers, *Fearless on Everest*, 196).

28 4/24   The Dzong Pen of Kharta Shikar is mentioned in the diary entries for 9–12 6/22.

30 4/24–2 5/24   Camps I and II were established on the East Rongbuk glacier in the same locations as in 1922 by parties of porters led by the Gurkha NCOs.

30 4/24   The female porter who carried a load plus her 3-year old child to Camp II appears in Noel's 1924 film *The Epic of Everest*.

2 5/24   Harke was one of the Gurkha NCOs. His note reported that the porters had successfully established Camp II and were being sent down again (*Everest 1924*, 57).

2 and 5 5/24   Hari Singh Thapa was a Gurkha member of the Survey of India who was attached to the expedition. He surveyed some of the areas to the north of Everest while the climbers attacked the mountain.

3 5/24   The Chunji La (or Chongay La) was an agent of the Shekar Dzong Pen, whose jurisdiction extended as far as the Rongbuk valley (*Everest 1922*, 42, 53, 57).

4 5/24   Umar was one of the Gurkha NCOs.

5 5/24   In Hingston's absence, Somervell, himself a medical man, acted as the expedition's medical officer.

11 5/24   Tam Ding was a porter who had acted as Somervell's servant during the journey across Tibet. Dorjay Pasang was one of the porters who went on to earn the title 'Tiger' for his exploits at high altitude.

12–14 5/24   Man Bahadur, a cobbler, and Sangloo were porters. Lance-Naik Shamshar, like Harke, was one of the Gurkha NCOs.

13 5/24   'No game was ever worth a rap ...' This couplet by George Whyte-Melville was a favourite of Norton's, and was quoted by him in his *Times* communiqué of 26 May, published on 16 June.

15 5/24   The ceremony at the Rongbuk Monastery features in the Noel film, *The Epic of Everest*.

20 5/24   Llakpa Tsering was a porter who had twice carried loads up to Camp V at about 25,000ft during the 1922 expedition.

23–24 5/24   Phu the cook (also spelt Pu or Poo) took part in all three Everest expeditions. Both he and Namgya had served at times as Somervell's servant, a fact which may have contributed to their successful rescue by Somervell.

24 5/24   The culmination of the rescue mission, as Somervell coaxed the four porters across the traverse, was memorably captured on film from a great distance by Noel in *The Epic of Everest*.

28–29 5/24   The most difficult part of the ascent to the North Col was a 150ft-high ice chimney (see diary entry for 20 5/24). The rope ladder improvised by Irvine greatly facilitated the carrying of loads to the North Col, and can be seen in a sequence in *The Epic of Everest*.

31 5/24–3 6/24   Two summit assaults were launched from Camp III on consecutive days. After spending the night of 31 May at Camp IV, Mallory and Geoff Bruce established Camp V at over 25,000ft on 1 June (not 2 June, as stated in *Everest 1924*, 96). Unable to persuade their porters (including Dorjay Pasang) to go any further, they returned to Camp IV on 2 June, passing Norton and Somervell who were on their way up to Camp V. The following day Norton and Somervell established Camp VI with the help of the four Tigers named in the diary entry. The site of Camp VI was rediscovered by the 2001 Mallory and Irvine Research Expedition at a height said to be around 26,700ft (100ft lower than estimated in 1924). Among the objects recovered was one of Norton's socks (Hemmleb and Simonson, *Detectives on Everest*, 101–13).

3 6/24   The name of the Tiger, given in the diary as Nuboo Yishay, is spelt Narbu Yishe in *Everest 1924* and in the subsequent literature.

4 6/24   The brief diary entry for this day provides the first, immediate record of Norton and Somervell's record-breaking summit attempt without supplementary oxygen.

6 6/24   Hingston climbed the steep ascent from Camp III to the North Col to examine Norton's eyes, in spite of the fact that 'he

had never previously climbed a mountain in the Alpine sense' (*Everest 1924*, 118). Noel's film, *The Epic of Everest*, shows Norton being carried into Camp III stone blind on the back of a porter, and groping his way into a tent.

7 6/24   The diary summarises Mallory's famous last message and note to Noel.

8–9 6/24   The entries succinctly convey Norton's anxiety and frustration as he waited for news of Mallory and Irvine. They also record his conviction, from the very start, that they had not been benighted on the mountain but had fallen off; and they summarise Odell's glimpse of the pair 'going strong' about the final step before the pyramid (now known as the 'second step').

10 6/24   Daily diary entries resume.

12–15 6/24   During these last few days Norton, as expedition leader, was exceptionally busy. Among other things, on 12 June he convened an expedition meeting to discuss the loss of Mallory and Irvine. All except Odell believed they had fallen from the mountain; Odell alone believed they had been benighted above Camp VI. He also sent a cable announcing the disaster, followed by longer communications, and wrote the letters of condolence to Mallory's and Irvine's families printed on pp. 130–2.

The 'monument' is the memorial cairn constructed above Base Camp which records the names of those who died on the three Everest expeditions of 1921, 1922 and 1924.

16 6/24   Noel accompanied by some of the porters set off immediately to rejoin his wife in the Chumbi valley. Hazard went with Hari Singh and others to survey the West Rongbuk glacier. The rest of the party headed west for a period of much-needed recuperation in the Rongshar valley.

17–18 6/24   The Lamna La, Kyetrak and Kombu La are all marked thus on the preliminary map of the area published after the 1921 reconnaissance expedition, and on some later maps. Sola Kombu is the first major settlement across the border in Nepal.

20–21 6/24   Norton was less impressed by the Rongshar valley and its flora than he had been by the Kama valley and the area to the east of Everest which they had explored during the recuperation period in 1922.

27 6/24   From here on the diary entries all the way back to Darjeeling are very succinct.

8 7/24   Hazard had returned separately from surveying the West Rongbuk glacier and headed for the Tsang Po without waiting for the rest of the expedition to catch up with him.

16–17 7/24   Hazard met up with the others at Kampa Dzong. He, Shebbeare and Odell then headed south to cross directly into Sikkim, while the remainder of the expedition carried on round by Phari.

22–24 7/24   Noel's wife had stayed with the MacDonalds at Yatung during the expedition. In his memoirs, David MacDonald mentions the lunch 'à la Everest' laid on for his family in the expedition mess-tent. He also recalls that Mrs Noel had had a premonition that a tragedy had occurred on Everest involving a close friend (but not her husband); and that this turned out to have been on the very day that Mallory (a friend of hers from childhood) had died (Macdonald, *Twenty Years in Tibet*, 300–1).

30–31 7/24   For Dr Graham and the Listers, see the entries for 26–28 3/24.

1 8/24   The expedition ended for Norton as it began, in a motor car, accompanied by General Bruce, who had remained in Darjeeling.

THREE LETTERS FROM THE 1924 EXPEDITION

Two Letters of Condolence

The letter to Irvine's father is taken from a photocopy of the original kindly supplied by Julie Summers, Irvine's biographer, who printed it in *Fearless on Everest*, 249–50. The letter to Mallory's widow, Ruth, is taken from a photocopy of the original kindly supplied by Mallory's son, the late John Mallory.

Norton's public tributes to Mallory and Irvine were printed in the communiqué for *The Times* dated 14 June, published on 5 July; in the reports of the public meeting of 17 October 1924 published in *The Alpine Journal*, 1924, 249–51 and *The Geographical Journal*, 1924, 441–3; and in *The Fight for Everest, 1924*, 145–9. These sources also contain his public statements as to the probable cause of their deaths, the likelihood of their having reached the summit, and Mallory's views on never risking lives in order to reach the top.

Simla, a hill-station in northern India, was the summer residence of the British Raj.

Letter to Sir Francis Younghusband

The original letter is preserved among the Younghusband papers in the British Library (Ms Eur. F197).

Norton's last *Times* communiqué, sent from Yatung on 22 July and published on 1 August, was followed by a final communiqué sent by General Bruce from Darjeeling on 14 August.

'My grandfather …', i.e. Alfred Wills, the pioneering Alpinist and builder of the family chalet above Sixt.

'One misconception …' The *Times* communiqué of 8 June, published on 26 June, contained an account of Norton and Somervell's summit attempt, written by Somervell. As the letter says, the printed article contained an unfortunate mistake, stating that Norton had only climbed 8ft higher than Somervell in an hour, instead of 80ft. The error was perpetuated when the communiqué was reprinted in *The Geographical Journal,* 1924, 158, but was corrected in the reprint in *The Alpine Journal*, 1924, 214 (see Further Reading for full details). The letter contains interesting details about Norton's solo climb above 28,000ft which, as he says, differ from Somervell's account in the *Times* communiqué.

The SS *Narkhunda* was an ocean liner built in 1920.

The reference to the Albert Hall is to the public meeting which was held on 17 October 1924 and which was already being planned before the end of the expedition.

AFTER EVEREST

❖ The Shipton quotations are from his posthumous appreciation, 'Norton of Everest', published in *The Geographical Journal*, vol. 121, no. 1 (March 1955), 84–5.

❖ Norton's Everest publications are listed in full below, Further Reading.

❖ Odell's words are cited by Hoyland, *Last Hours on Everest*, 68.

❖ The Mallory quote comes from his letter of 19 April 1924, cited in Robertson, *George Mallory*, 231.

❖ Somervell's assessment is found in a letter of condolence he wrote to Norton's widow in 1954.

❖ Younghusband quote from *Everest, The Challenge*, 59–60.

❖ The meeting with Ang Tsering is described in Band, *Everest: 50 Years on Top of the World*, 244–5.

❖ Letter from John Hunt in the Norton Everest Archive.

GLOSSARY

A and N rations	rations from the Army and Navy stores
A.1.	first class
accaulis	type of gentian, Latin name *gentiana acaulis*
ammon	Himalayan great wild sheep with large horns
aneroid	barometer without liquid
arête	ridge
arrak	locally produced spirit
atta	wheat flour
babu	Indian term for clerk, official
bagh	garden, orchard
bandobast, bundobust	arrangement, business, organisation
basti	village, shanty town
bergschrund	large crevasse at head of glacier, formed by its break from main mass of mountain
Bhotias, Bhutias	Tibetan mountain people
bundobust	*see* bandobast
burhel	Himalayan wild sheep
champa, sampa, tsampa, tsampha, tzampha	Tibetan barley dish
chang	local beer
chapatti	flat loaf of unleavened bread
chit	short letter, note
cho(mo)	goddess (a common prefix for mountain names)
chu	stream, water
chukhor	Indian partridge
col	saddle between two mountains
comforter	long scarf
cooly	porter
couloir	steep gully
dak	post, postal service
dak bungalow	staging-post for the postal service, also used by travellers to spend the night, hence rest house
dhokpa	encampment of nomad shepherds
dis aliter visum	the gods decreed otherwise
Drury Lane	street in London known for its theatres; hence theatrical

dzong, jong	fortress
dzong pen, jong pen	senior official in charge of dzong
finneskoe	soft fur-lined boot
flea bag	sleeping bag
gampa, gompa	monastery
gaudy	happy, well
gembu, gyembu	official subordinate to dzong pen; headman
gendarme	pinnacle, usually of rock, on a mountain ridge
ghil	probably jheel, artificial lake, area of standing water
gompa	*see* gampa
Gurkhali	language spoken by the Gurkhas
Gurkhas	Nepalese hill people, many of whom served in the army
gyembu	*see* gembu
hoosh	thick soup
hum mani pedmi hum	Buddhist mantra, *see* om mani padme hum
jong	*see* dzong
jong pen, JP	*see* dzong pen
kad	kudh, precipitous slope, ravine
Khaskura	language of the Gurkhas
khubber	news, especially about game
kyang	Tibetan wild ass
la	mountain pass
lama	monk
lance-naik	equivalent of lance-corporal
ling	place, garden

Maconochi	tinned stew
mani wall	wall incorporating stones inscribed with the mantra om mani padme hum
marg	mountain meadow
Meade tent	two- or three-man tent, named after the manufacturer
meta	type of solid fuel in tablet form
moraine	line of debris on or left by a glacier
more suo	in his own way, in his usual way (Latin)
Mummery tent	high-altitude tent, named after the Alpine pioneer
nullah	ravine, water-course, often used of dried-up water-courses
om mani padme hum	Buddhist mantra: Hail, jewel of the lotus flower
Parri wallah	man from Phari
Primus	liquid-fuel stove
pro tem.	for the time being (Latin: *pro tempore*)
puttoo	coarse goats-wool fabric made in Cashmere
ri	mountain
roarer cooker	a heavy-duty cooker weighing 40lbs, so-called from the noise it made
rs	rupees
sampha	*see* champa
sangar	small, square, roofless enclosure with stone walls about 3ft high
serac	pillar of ice on a glacier
shegar, shekar	local head-man
Sherpas	Nepalese hill people
shikar	hunting, sport

shikarri	hunter	**tse**	peak, summit
supercilium	eyebrow, mark over eye	**tsomo**	lake
syce	groom		
		wallah	official, individual with specific job
tamasha, tomasha	entertainment, show	**Whymper tent**	four-man tent, named after the famous Alpine pioneer
tiffin	light snack		
tomasha	*see* tamasha		
trak	cliff, rock	**x**	abbreviation for yard(s)
ts	tankas (unit of Tibetan currency)		
tsampa, tsampha	*see* champa	**zoh**	cross between a yak and a cow

Further Reading

1. THE EVEREST PUBLICATIONS OF E.F. NORTON

Norton was to have collaborated with Longstaff in the chapter on natural history in the 1922 expedition book. In particular, he was to have written a section on botany. However, he was prevented from contributing because of his duties on active service in the Dardanelles. Consequently, he left no published account of any aspect of the 1922 expedition.

The situation with the 1924 expedition was very different. When he took over the leadership of the expedition following General Bruce's withdrawal through ill health, he inherited the responsibility for penning a series of dispatches to *The Times*. General Bruce had sent the first two dispatches, so Norton's dispatches are numbered 3 onwards. These constitute his first public narrative of the expedition, and contain some vivid details and turns of phrase which were not repeated in the subsequent, more considered accounts. Some of the dispatches were partly written by other members of the expedition. They were published in *The Times* two or three weeks after they were written, as follows:

3. 14 April, Khamba Dzong: published 29 April.
4. 22 April, Kyishong: published 12 May.
5. 29 April, Base Camp, Rongbuk Glacier: published 17 May.
6. 13 May, Base Camp, Rongbuk: published 31 May.
7. 26 May, Camp no 1, East Rongbuk glacier: published 16 June.
8. 8 June, with addendum dated 11 June, Camp Three, East Rongbuk glacier: published 26 June.
9. 14 June, Rongbuk Base Camp: published 5 July.
    19 June, telegram announcing the death of Mallory and Irvine, dispatched from Phari Dzong: published 21 June.
10. 19 June, Kyetrak: published 15 July.
11. 24 June, Rongshar valley: published 17 July.
12. 5 July, Tingri Dzong: published 25 July.
13. 22 July, Yatung: published 1 August.

General Bruce completed the series with a final dispatch from Darjeeling on 14 August announcing the return of the expedition a fortnight earlier.

Dispatches 6–9 were reprinted (incorporating a few corrections from a comparison of the original cablegrams with the texts printed in *The Times*) as 'The Mount Everest Dispatches' in *The Geographical Journal*, vol. 64, no. 2 (August 1924), 145–65, and nos. 6–13 were reprinted as 'The Mount Everest Dispatches' in *The Alpine Journal*, vol. 36, no. 229 (November 1924), 196–241.

On 17 October 1924 a joint meeting of the Alpine Club and the Royal Geographical Society was held at the Royal Albert Hall, on the same day that a memorial service for Mallory and Irvine was held at St Paul's Cathedral. The story of the 1924 expedition was recounted by General Bruce, Norton, Geoff Bruce and Odell, and the texts of their addresses were published in the respective journals of the two societies. Norton's contributions consisted of an appreciation of the individual members of the expedition and an account of his own attempt on the summit with Somervell. They were printed as 'The personnel of the expedition' and 'The climb with Mr. Somervell to 28,000 feet', in *The Alpine Journal*, vol. 36, no. 229 (November 1924), 244–51 and 260–5 and in *The Geographical Journal*, vol. 64, no. 6 (December 1924), 436–43 and 451–5.

On 15 December 1924, Norton gave a paper to the Alpine Club reflecting on the lessons to be learnt from the 1924 expedition and the prospects for future attempts. It was published as 'The Problem of Mt. Everest', *The Alpine Journal*, vol. 37, no. 230 (May 1925), 1–22. Later in 1925 appeared the official book of the expedition: E.F. Norton, *The Fight for Everest: 1924* (London: Edward Arnold and Co., 1925). As with the two previous expedition books sponsored by the Mount Everest Committee, it was a collaborative effort whose title-page bore the name of the lead author, in this case Norton. He contributed the chapters on 'The March across Tibet' (31–53), 'The North Col' (73–98), 'Norton and Somervell's attempt' (99–119), 'The Return to Base Camp' (144–54) and 'Future Possibilities' (193–204), this last being based on his article in *The Alpine Journal*. He also contributed a note on 'Preliminary Organization, Journey and March' (338–42).

In 1950, he published another article reviewing the experiences of the pre-war expeditions and re-affirming his belief that the summit could be reached: 'Mount Everest: The Last Lap', *The Alpine Journal*, vol. 57, no. 280 (May 1950), 285–92.

Finally, in 1954, shortly before his death, he published a review of the account of the successful 1953 Everest expedition by John Hunt, *The Ascent of Everest* (London, 1953) in *The Geographical Journal*, vol. 120, part 1 (March 1954), 82–3.

2. SELECT LIST OF PUBLICATIONS ON THE EARLY EXPEDITIONS

The essential works of reference on the early Everest expeditions are the three official books, namely:

C.K. Howard-Bury, *Mount Everest: The Reconnaissance, 1921* (London, 1922).
C.G. Bruce, *The Assault on Mount Everest, 1922* (London, 1923).
E.F. Norton, *The Fight for Everest: 1924* (London, 1925).

The most useful additional literature by those who participated or by later authors is as follows:

G.I. Finch, *The Making of a Mountaineer* (London, 1924).
F. Younghusband, *The Epic of Mount Everest* (London, 1926).
J.B.L. Noel, *The Story of Everest* (New York, 1927), reprinted as *Through Tibet to Everest* (London, 1989).
D. Pye, *George Leigh Mallory: A Memoir* (Oxford, 1927), reprinted with additions (Bangkok, 2002).
D. Macdonald, *Twenty Years in Tibet* (London, 1932).
F. Younghusband, *Everest, The Challenge* (London, 1936).
T.H. Somervell, *After Everest* (London, 1936).
J.R. Ullman, *Kingdom of Adventure: Everest* (London, 1948).
E. Shipton, 'Norton of Everest', *The Geographical Journal*, vol. 121, no. 1 (March 1955), 84–5.
D. Robertson, *George Mallory* (London, 1969), reprinted with additions (London, 1999).
H. Carr, *The Irvine Diaries: Andrew Irvine and the Enigma of Everest 1924* (Goring, 1979).
T. Holzel and A. Salkeld, *The Mystery of Mallory and Irvine* (London, 1986), revised and expanded edition (London, 1999).
C. Howard-Bury and G. Leigh Mallory, ed. M. Keaney, *Everest Reconnaissance: The First Expedition of 1921* (London, 1991), a partial reprint of Howard-Bury's book of the 1921 expedition with additional material.

J. Hemmleb, L.A. Johnson and E.R. Simonson, *Ghosts of Everest* (London, 1999).

D. Breashears and A. Salkeld, *Last Climb – The Legendary Everest Expeditions of George Mallory* (Washington, 1999).

J. Summers, *Fearless on Everest: The Quest for Sandy Irvine* (London, 2000).

P. and L. Gillman, *The Wildest Dream: Mallory, His Life and Conflicting Passions* (London, 2000).

J. Hemmleb and E. Simonson, *Detectives on Everest – The 2001 Mallory and Irvine Research Expedition* (Seattle, 2002).

S. Noel, *Everest Pioneer: The Photographs of Captain John Noel* (Stroud, 2003).

Royal Geographical Society, *Everest, Summit of Achievement* (London, 2003).

G. Band, *Everest: 50 Years on Top of the World* (London, 2003).

R. Macfarlane, *Mountains of the Mind* (New York, 2003).

P. Gillman, ed., *Climbing Everest, by George Leigh Mallory* (London, 2010).

W. Davis, *Into the Silence: The Great War, Mallory and the Conquest of Everest* (London, 2011), with very full bibliography.

G. Hoyland, *Last Hours on Everest – The Gripping Story of Mallory and Irvine's Fatal Ascent* (London, 2013).

Also of great interest are the films of the 1922 and 1924 expeditions by John Noel, *Climbing Mount Everest* and *The Epic of Everest: The Immortal Film Record of this Historic Expedition*. The latter was re-released by the British Film Institute in 2013.

Index

Visit our website and discover thousands of other History Press books.

www.thehistorypress.co.uk

The History Press